"An enthralling reconstruction o[...]
−*NEW YORK TIMES BOOK*[...]

"A meaty, exceedingly well-researched a[...]
dramatic readability *Wittgenstein's Poke*[...]
Winchester's *The Professor and the Madman;* in the depth[...]
breadth of its scholarship it evokes Carl Schorske's *Fin-de-Siecle
Vienna* . . . a marvel of passionate journalism."
−*SAN FRANCISCO CHRONICLE*

"Edmonds and Eidinow['s] account of Wittgenstein's
notoriously difficult ideas is admirably clear."
−*TIME*

"An excellent piece of philosophical journalism . . .
taking us easily and pleasurably through Viennese history
and culture, Cambridge philosophy, and the personal
quirks of the protagonists."
−*FINANCIAL TIMES*

"Edmonds and Eidinow go to fascinating lengths to
provide background for this altercation . . . a remarkably lucid
overview of twentieth-century philosophy."
−*NEWSDAY*

"A wonderful yarn . . . For all his superhuman austerity,
Wittgenstein (like the great physicist Niels Bohr) was fond of
relaxing with cowboy movies and detective stories.
Wittgenstein's Poker is just the thing for us mentally
flabbier mortals in search of a snack."
−*BOSTON GLOBE*

"One of the year's most entertaining and intellectually rich
books. . . . Tightly constructed and extraordinarily well written,
this is a marvelous blend of lay and
academic scholarship. . . . A classic of its kind."
−*PUBLISHERS WEEKLY* (STARRED REVIEW)

DAVID EDMONDS and JOHN EIDINOW are award-winning journalists with the BBC. This book, their first, has been translated into more than a dozen languages.

Ludwig Wittgenstein

Karl Popper

WITTGENSTEIN'S POKER

*The Story of a Ten-Minute Argument
Between Two Great Philosophers*

David Edmonds and John Eidinow

An Imprint of HarperCollins Publishers

To Hannah and Herbert Edmonds
and to Elisabeth Eidinow

I know that queer things happen in this world. It's one of the few things I've really learned in my life.

— WITTGENSTEIN

Great men can make great mistakes.

— POPPER

Contents

WITTGENSTEIN'S POKER

King's College, seen from the river. The scene of the poker incident, room H3, is on the left on the first floor.

—◆◆◆—

The Poker

History is affected by discoveries we will make in the future.

—POPPER

ON THE EVENING OF FRIDAY, 25 October 1946 the Cambridge Moral Science Club—a weekly discussion group for the university's philosophers and philosophy students—held one of its regular meetings. As usual, the members assembled in King's College at 8:30, in a set of rooms in the Gibbs Building—number 3 on staircase H.

That evening the guest speaker was Dr. Karl Popper, down from London to deliver an innocuous-sounding paper, "Are There Philosophical Problems?" Among his audience was the chairman of the club, Professor Ludwig Wittgenstein, considered by many to be the most brilliant philosopher of his time. Also present was Bertrand Russell, who for decades had been a household name as a philosopher and radical campaigner.

Popper had recently been appointed to the position of Reader in Logic and Scientific Method at the London School of Economics (LSE). He came from an Austrian-Jewish background and was newly arrived in Britain, having spent the war years lecturing in New Zealand. *The Open Society and Its Enemies*, his remorseless demolition of totalitarianism, which he had begun on the day Nazi troops entered Austria and completed as the tide of war turned, had just been published in England. It had immediately won him a select group of admirers—among them Bertrand Russell.

This was the only time these three great philosophers—Russell, Wittgenstein, and Popper—were together. Yet, to this day, no one can agree precisely about what took place. What is clear is that there were vehement exchanges between Popper and Wittgenstein over the fundamental nature of philosophy—whether there were indeed philosophical problems (Popper) or merely puzzles (Wittgenstein). These exchanges instantly became the stuff of legend. An early version of events had Popper and Wittgenstein battling for supremacy with red-hot pokers. As Popper himself later recollected, "In a surprisingly short time I received a letter from New Zealand asking if it was true that Wittgenstein and I had come to blows, both armed with pokers."

Those ten or so minutes on 25 October 1946 still provoke bitter disagreement. Above all, one dispute remains heatedly alive: did Karl Popper later publish an untrue version of what happened? Did he lie?

If he did lie, it was no casual embellishing of the facts. If he lied, it directly concerned two ambitions central to his life: the defeat at a theoretical level of fashionable twentieth-century linguistic philosophy and triumph at a personal level over Wittgenstein, the sorcerer who had dogged his career.

Popper's account can be found in his intellectual autobiography, *Unended Quest*, published in 1974. According to this version of events, Popper put forward a series of what he insisted were real philosophical problems. Wittgenstein summarily dismissed them all. Popper recalled that Wittgenstein "had been nervously playing with the poker," which he used "like a conductor's baton to emphasize his assertions," and when a question came up about the status of ethics, Wittgenstein challenged him to give an example of a moral rule. "I replied: 'Not to threaten visiting lecturers with pokers.' Whereupon Wittgenstein, in a rage, threw the poker down and stormed out of the room, banging the door behind him."

When Popper died, in 1994, newspaper obituarists picked up his telling of the tale and repeated it word for word (including the wrong date for the meeting—the 26th, not the 25th). Then, some three years after Popper's death, a memoir published in the proceedings of one of Britain's most learned bodies, the British Academy, recounted essentially the same sequence of events. It brought down a storm of protest on the head of the author, Popper's successor at the LSE, Professor John Watkins, and sparked off an acerbic exchange of letters in the pages of the London *Times Literary Supplement*. A fervent Wittgenstein supporter who had taken part in the meeting, Professor Peter Geach, denounced Popper's account of the meeting as "false from beginning to end."

It was not the first time Professor Geach had made that allegation. A robust correspondence followed as other witnesses or later supporters of the protagonists piled into the fray.

There was a delightful irony in the conflicting testimonies. They had arisen between people all professionally concerned with theories of epistemology (the grounds of knowledge), understanding, and truth. Yet they concerned a sequence of events where those who disagreed were eyewitnesses on crucial questions of fact.

This tale has also gripped the imagination of many writers: no biography, philosophical account, or novel involving either man seems complete without a—frequently colorful—version. It has achieved the status, if not of an urban myth, then at least of an ivory-tower fable.

But why was there such anger over what took place more than half a century before, in a small room, at a regular meeting of an obscure university club, during an argument over an arcane topic? Memories of the evening had remained fresh through the decades, persisting not over a complex philosophical theory or a clash of ideologies, but over a quip and the waving—or otherwise—of a short metal rod.

WHAT DO THE INCIDENT and its aftermath tell us about Wittgenstein and Popper, their remarkable personalities, their relationship, and their beliefs? How significant was it that they both came from fin de siècle Vienna, both born into assimilated Jewish families, but with a great gulf of wealth and influence between them? And what about the crux of the evening's debate: the philosophical divide?

Wittgenstein and Popper had a profound influence on the way

we address the fundamental issues of civilization, science, and culture. Between them, they made pivotal contributions both to age-old problems such as what we can be said to know, how we can make advances in our knowledge, and how we should be governed, and to contemporary puzzles about the limits of language and sense, and what lies beyond those limits. Each man believed that he had freed philosophy from the mistakes of its past, and that he carried responsibility for its future. Popper saw Wittgenstein as philosophy's ultimate enemy. Yet the story of the poker goes beyond the characters and beliefs of the antagonists. It is inseparable from the story of their times, opening a window on the tumultuous and tragic history that shaped their lives and brought them together in Cambridge. And it is the story of the schism in twentieth-century philosophy over the significance of language: a division between those who diagnosed traditional philosophical problems as purely linguistic entanglements and those who believed that these problems transcended language. In the end, of course, it is the story of a linguistic puzzle in itself: to whom did Popper utter what words in that room full of witnesses, and why?

Before we begin to delve into the personalities, the history, and the philosophy of those ten minutes in H3, let us introduce what are fixed and ascertainable: the place, the witnesses, and their avowed recollections.

2

Memories Are Made of This

*Memory: "I see us still, sitting at that table." But have I
really the same visual image — or one of those that I had
then? Do I also certainly see the table and my friend from
the same point of view as then, and not see myself?*

—WITTGENSTEIN

THE GIBBS BUILDING OF KING'S COLLEGE IS a mas-
sive, severely classical block, constructed of white Portland stone.
It was designed in 1723 by James Gibbs, who was the college's second choice: the initial plan, by Nicholas Hawksmoor, one of the
premier architects of the day, was too expensive, and the building's remarkable and much praised restraint in decoration was the
result of King's being short of money.

Viewed from the street, King's Parade, H3 is on the right-hand side of the building, on the first floor. The echoing approach, up a flight of uncarpeted wooden stairs past bare walls, is chill and uninviting. The double front door leads directly into the sitting room. Two long windows with window seats overlook the spacious elegance of the college front court and, filling the view to the left, Henry VI's great limestone chapel, a supreme example of perpendicular architecture. In the silence of an October evening, the singing of King's celebrated choir will break in on donnish concentration.

The feature of H3 at the heart of this decades-long quarrel, the fireplace, is enclosed by a marble surround above which is a carved wooden mantelpiece. It is a small, black, iron affair—more *The Road to Wigan Pier* than *Brideshead Revisited*. To its right are the doors to two smaller rooms. With views over the big lawn sweeping down to the river Cam, at the time of the meeting they were a study and a bedroom, although the bedroom has since been turned into a second study. In those days and for some years after, most members of Cambridge colleges—undergraduates and fellows alike—were expected to dart in their dressing gowns across the courts to a communal bathroom.

In 1946 the splendor of the Gibbs Building's exterior was not reflected in the state of its rooms. This was barely a year after the end of the Second World War. Blackout curtains were still hanging—a reminder of the Luftwaffe's recent threat. Paintwork was chipped and grimy, the walls in urgent need of a wash. Although its tenant was a don, Richard Braithwaite, H3 was just as neglected as the other rooms in the building, squalid, dusty, and dirty. Heating was dependent on open fires—central heating and baths were not installed until after the ultrasevere winter of 1947, when even the water that collected in the gas pipes froze, blocking them—and the inhabitants protected their clothes with their gowns when humping sacks of coal.

NORMALLY, despite the eminence of many of the speakers, only fifteen or so people would turn up to the Moral Science Club; significantly, for Dr. Popper there were perhaps double that number. The medley of undergraduates, graduates, and dons squeezed into whatever space they could find. Most of those who had been at Wittgenstein's late-afternoon seminar, held in his barely furnished rooms at the top of a tower of Whewell's Court— across the street from the great gate of Trinity College, where he held a fellowship—rejoined him in King's.

Conducted twice weekly, these seminars offered students a mesmerizing experience. As Wittgenstein struggled with a thought there would be a long moment of agonized silence; then, when the thought was formed, a sudden burst of ferocious energy. Permission was granted for students to attend—but on condition that they were not there merely as "tourists." On the afternoon of 25 October an Indian graduate, Kanti Shah, took notes. What did it mean, Wittgenstein wanted to know, to speak to oneself? "Is this something fainter than speaking? Is it like comparing 2+2=4 on dirty paper with 2+2=4 on clean paper?" One student suggested a comparison with a "bell dying away so that one doesn't know if one imagines or hears it." Wittgenstein was unimpressed.

Meanwhile, in Trinity College itself, in a room once occupied by Sir Isaac Newton, Popper and Russell were drinking China tea with lemon and eating biscuits. On this chilly day, both would have had reason to be grateful for the draft excluders newly placed around the windows. It is not known what they talked about, though one account has them plotting against Wittgenstein.

HAPPILY, philosophy appears good for longevity: of the thirty present that night, nine, now in their seventies or eighties, responded by letter, phone and, above all, e-mail from across the globe—from England, France, Austria, the United States, and New Zealand—to appeals for memories of that evening. Their ranks include a former English High Court judge, Sir John Vinelott, famous both for the quiet voice with which he spoke in court and for the sharpness with which he responded to counsel who asked him to speak up. There are five professors. Professor Peter Munz had come to St. John's from New Zealand and returned home to become a notable academic. His book *Our Knowledge of the Search for Knowledge* opened with the poker incident: it was, he wrote, a "symbolic and in hindsight prophetic" watershed in twentieth-century philosophy.

Professor Stephen Toulmin is an eminent philosopher of widely ranging interests who spent the latter part of his academic career teaching at universities in the United States. He wrote such leading works as *The Uses of Argument*, and is coauthor of a demanding revisionist text on Wittgenstein, placing his philosophy in the context of Viennese culture and fin de siècle intellectual ferment. As a young King's research fellow, he turned down a post as assistant to Karl Popper.

Professor Peter Geach, an authority on logic and the German logician Gottlob Frege (among many other things), lectured at the University of Birmingham, and then at Leeds. Professor Michael Wolff specialized in Victorian England, and his academic career took him to posts at Indiana University and the University of Massachusetts. Professor Georg Kreisel, a brilliant mathematician, taught at Stanford; Wittgenstein had declared him the most able philosopher he had ever met who was also a mathematician. Peter Gray-Lucas became an academic and then switched to business, first in steel, then photographic film, then papermaking. Stephen Plaister, who was married in the freezing winter of 1947, became a prep-school master, teaching classics.

Wasfi Hijab deserves a special mention. He was the secretary of the Moral Science Club at the time of the fateful meeting. No real prestige was attached to the position, he says. He cannot even remember how he came to hold it—probably a case of Buggin's turn. His job as secretary was to fix the agenda for the term, which he would do after consulting with members of the faculty. In his period of office he succeeded in persuading not just Popper to travel to Cambridge, but also the man who brought the news of logical positivism from Vienna to England, A. J. Ayer. Ayer always found it an "ordeal" to speak in front of Wittgenstein, but nevertheless replied to Hijab's invitation by saying that he would gladly talk to the society, even though in his opinion "Cambridge philosophy was rich in technique but poor in substance." "That," says Hijab, "shows how much he knew."

Hijab's Cambridge experience says much about Wittgenstein. He had arrived in Cambridge in 1945 on a scholarship from Jerusalem, where he had taught mathematics in a secondary school. His goal was to switch disciplines by studying for a doctorate in philosophy. Three years later he left with his Ph.D. unfinished. He had made a mistake fatal to his ambitions: against all advice—from Richard Braithwaite among others—he had asked Wittgenstein to be his supervisor. To general astonishment, Wittgenstein had agreed.

Hijab remembers his tutorials well. They were, when weather

permitted, ambulatory. Round and round the manicured Trinity fellows' garden they would walk: he, Wittgenstein, and a fellow student, Elizabeth Anscombe, deep in discussion of the philosophy of religion. "If you want to know whether a man is religious, don't ask him, observe him," said Wittgenstein. In his supervisor's presence Hijab was mostly struck dumb with sheer terror; in his absence, he says, he sometimes demonstrated he was catching a spark off the old master.

Wittgenstein, Hijab now reflects, destroyed his intellectual foundations, his religious faith, and his powers of abstract thought. The doctorate abandoned, for many years after leaving Cambridge he put all thought of philosophy aside and took up mathematics again. Wittgenstein, he says, was "like an atomic bomb, a tornado—people just don't appreciate that."

Nevertheless, Hijab retains that fierce loyalty to his teacher that Wittgenstein could inspire. "People often say that all philosophy is just a footnote to Plato," Hijab says, "but they should add, 'until Wittgenstein.'" His devotion finally had its reward. In 1999 he caused a sensation at a Wittgenstein conference in Austria when he more or less gate-crashed the program, but was then given two extra sessions for his discourses on the master, meriting a write-up in the ultraserious *Neue Zürcher Zeitung*. From Austria, Hijab moved on to the Wittgenstein Archive in Cambridge, to hold seminars there. It took him, he said, half a century to recover from his "overexposure" to Wittgenstein. Now he wanted to make up for lost time.

For the full story of the confrontation between Wittgenstein and Popper we must wait until all the evidence is in. But there can be no better place to start than with our eyewitnesses.

Conventionally, we should feel a chill in the air as a glance around the room summons up the ghostly crowd waiting for Dr. Popper to begin his paper and picks out our nine, now returned to youth, from among them. Inevitably the eye goes first to the crowning intellects of the evening. In front of the fireplace, placidly smoking his pipe, is the silver-haired Bertrand Russell. To Russell's left, facing the audience, is an apparently quiet and insignificant figure, Karl Popper. One or two of the undergraduates are noting his prominent ears—altogether out of proportion to his small stature—to joke about over a pint after the meeting. Popper is taking the measure of his adversary, whom he has thought about so much but until now never actually seen: Wittgenstein, the club's chairman, sitting to Russell's right. He too is small, but filled with nervous energy, passing his hand over his forehead as he waits to open the meeting and looking at Popper with those penetrating blue eyes and their "intensely white and large surrounds that make you feel uncomfortable."

Wittgenstein and Popper are our reason for being here. But now the eye moves on to the young Palestinian graduate, Wasfi Hijab. He is clutching the Moral Science Club minute book, in which he will later pen the understatement describing the confrontation of the evening: "The meeting was unusually charged."

It was Hijab who had sent the neatly handwritten invitation to Popper and negotiated a change of date, from the club's habitual Thursday to Friday, to suit the guest. Like all such secretaries, he feels responsible for the guest's showing up and frets about his arrival until he actually sees him in the flesh. Popper's firm handshake is an early sign that his slight frame conceals an assertive personality.

Sitting nearby is one of Popper's closest friends in Cambridge,

Peter Munz, researching for a postgraduate degree in history. Munz is one of only two to have studied under both Wittgenstein and Popper: he was taught by Popper in New Zealand during the war and, as a transparently earnest, bright student, just a few weeks earlier had been welcomed by Wittgenstein into his Whewell's Court seminars. Munz recalls Popper pacing slowly across the room, throwing and catching a piece of chalk, never once breaking stride, and speaking in long and perfectly constructed sentences. Now he has encountered Wittgenstein, who wrestles visibly with his ideas, holding his head in his hands, occasionally throwing out staccato remarks, as though each word were as painful as plucking thorns, and muttering, "God I am stupid today" or shouting, "Damn my bloody soul! . . . Help me someone!"

Peter Munz in 1946, having survived the intellectual rigors of being taught by both Popper and Wittgenstein.

Then, there is John Vinelott, at twenty-three his features still showing the strain of his recent naval service in the Far East. A chance happening during the war brought him to this spot. Before joining the navy, he was a student of languages at London University. Then, browsing in a bookshop in Colombo, the capital of what he knew as Ceylon, now Sri Lanka, he picked up a copy of Wittgenstein's *Tractatus Logico-Philosophicus* and was immediately riveted. At the war's end he switched to Cambridge "to sit at Wittgenstein's feet." The skeptical eyes that will later disconcert so many litigants and barristers are now weighing up the guest speaker, Popper. That afternoon's session in Whewell's Court had been an even more vigorous intellectual workout than usual. Besides the puzzle of speaking to oneself, they had discussed the flexibility of the rules of mathematics. "Suppose you had all done arithmetic within this room only," Wittgenstein had hypothesized. "And suppose you go into the next room. Mightn't this make 2+2=5 legitimate?" He had pushed this apparent absurdity further. "If you came back from the next room with 20×20=600, and I said that was wrong, couldn't you say, 'But it wasn't wrong in the other room.'" Vinelott is still preoccupied with this. He has never before met a man of such intensity: "incandescent with intellectual passion" will be his memory.

Near the front sits a Wittgenstein ultra: Peter Geach, a postgraduate, though currently without any official Cambridge raison d'être. However, his wife, Elizabeth Anscombe, is a graduate student at the women's college, Newnham, and, like her husband, a member of the MSC. Tonight she is at home in Fitzwilliam Street, just beyond King's Parade, looking after their two young children. Both husband and wife are very close to Wittgenstein: she will become one of his heirs, translators, and literary executors, and a leading philosopher in her own right. Wittgenstein refers to her fondly as "old man." A near-contemporary description of her is "stocky . . . wearing slacks and a man's jacket." Together Elizabeth and Peter make a formidable academic couple, both with first-class degrees in what is said to be Oxford's toughest intellectual challenge, *Literae Humaniores*, the study of ancient Greek and Latin literature, Greek and Roman history, and ancient and modern philosophy. Their philosophy is informed by an unwavering commitment to Roman Catholicism. In Peter's case this may in part be a reaction to the fickleness of his father, who was in the habit of switching between religions every few months without apparent agonies of conscience; in Elizabeth's case to her being a convert.

Searching among the expectant crowd we can also locate Stephen Toulmin, Peter Gray-Lucas, Stephen Plaister, and Georg Kreisel. All four have come to Cambridge after contributing to the war effort. Originally a student of mathematics and physics, Toulmin had been based in a radar research station. Now, at twenty-four, having given up science, he is a graduate student in philosophy: his doctoral thesis is considered of such a high standard that it will be taken for publication by Cambridge University Press even before being accepted by the examiners. He has rushed here from the cottage which he rents from G. E. Moore, the former Professor of Philosophy, and which is at the end of Moore's garden. Peter Gray-Lucas, a talented linguist, fluent in German, played his part in the war at the top-secret decoding center at Bletchley Park, where so much of the Nazis' fighting strategy was undone. Georg Kreisel, Jewish and Austrian-born, was in the Admiralty; he is one of the few people not to be intimidated or overawed by Wittgenstein. Kreisel delights in the cruder of

Wittgenstein's ceaseless stream of aphorisms, such as "Don't try and shit higher than your arse," which Wittgenstein applied to philosophers like Popper who thought they could change the world. Stephen Plaister is less engaged with philosophy and has little contact with Wittgenstein. He will, however, always treasure one memory. After bumping into Wittgenstein and Kreisel in the street, Plaister was later told by Kreisel that Wittgenstein had liked his face. And there, standing out among the ex-servicemen for his youthful demeanor, is the fresh-faced Michael Wolff, straight from school at nineteen, and feeling a bit out of his depth.

They and the rest of our spectral phalanx are for the most part dressed in heavy sports jackets, gray flannels, regimental or school ties, perhaps a waistcoat or Fair Isle pullover. Remnants of a service uniform can still be seen on those short of clothing coupons. One or two might have "I was there" suede desert boots and cavalry-twill trousers. Wittgenstein's disciples stand out instantly for their aping of the master: casual, even sloppy, in open-neck shirts.

It is only to be expected that each of those present in that crowded room has a slightly different recollection of the night's events. Some had a restricted view. One thing happened on top of another, making the precise sequence uncertain. The flow of debate was so fast that it was difficult to follow. But most share one memory: the poker itself.

"Consider this poker," Peter Geach hears Wittgenstein demand of Popper, picking up the poker and using it in a philosophical example. But, as the discussion rages on between them, Wittgenstein is not reducing the guest to silence (the impact he is accustomed to), nor the guest silencing him (ditto). Finally, and only after having challenged assertion after assertion made by

Popper, Wittgenstein gives up. At some stage he must have risen to his feet, because Geach sees him walk back to his chair and sit down. He is still holding the poker in his hand. With a look of great exhaustion on his face, he leans back in his chair and stretches out his arm toward the fireplace. The poker drops on to the tiles of the hearth with a little rattle. At this point Geach's attention is caught by the host, Richard Braithwaite. Alarmed by Wittgenstein's gesticulating with the poker, he is making his way in a crouching position through the audience. He picks up the poker and somehow makes away with it. Shortly afterward Wittgenstein rises to his feet and, in a huff, quietly leaves the meeting, shutting the door behind him.

Michael Wolff sees that Wittgenstein has the poker idly in his hand and, as he stares at the fire, is fidgeting with it. Someone says something that visibly annoys Wittgenstein. By this time Russell has become involved. Wittgenstein and Russell are both standing. Wittgenstein says, "You misunderstand me, Russell. You always misunderstand me." He emphasizes "mis," and "Russell" comes out as "Hrussell." Russell says, "You're mixing things up, Wittgenstein. You always mix things up." Russell's voice sounds a bit shrill, quite unlike when lecturing.

Peter Munz watches Wittgenstein suddenly take the poker—red-hot—out of the fire and gesticulate with it angrily in front of Popper's face. Then Russell—who so far has not spoken a word—takes the pipe out of his mouth and says firmly, "Wittgenstein, put down that poker at once!" His voice is high-pitched and somewhat scratchy. Wittgenstein complies, then, after a short wait, gets up and walks out, slamming the door.

From where Peter Gray-Lucas is sitting, Wittgenstein seems to be growing very excited about what he obviously believes is Popper's improper behavior and is waving the poker about. Wittgenstein is acting in "his usual grotesquely arrogant, self-opinioned, rude and boorish manner. It made a good story afterward to say that he had 'threatened' Popper with a poker." Stephen Plaister, too, sees the poker raised. It really seems to him the only way to deal with Popper, and he has no feeling of surprise or shock.

To Stephen Toulmin, sitting only six feet away from Wittgenstein, nothing at all out of the ordinary is occurring; nothing that in hindsight would merit the term "incident." He is focusing on Popper's attack on the idea that philosophy is meaningless and his production of various examples. A question about causality arises, and at that point Wittgenstein picks up the poker to use as a tool in order to make a point about causation. Later in the meeting— but only after Wittgenstein has left—he hears Popper state his poker principle: that one should not threaten visiting lecturers with pokers.

There is also written testimony from Hiram McLendon, an American from Harvard, who spent the academic year 1946–47 in Cambridge studying under Russell and was there in H3. Such an impact did the evening have on him that many years later he wrote up his memories, checking his narrative with Russell, who approved it. The florid description casts his former tutor in the role of hero—"a towering giant, a roaring lion, a rod of reproof." Popper, he wrote, had delivered his paper almost with "apology for its boldness." It got a stormy reception, with those in the audience becoming increasingly agitated. Wittgenstein turned active, grabbed the iron poker, and waved it in a hostile manner, his voice rising in pitch, as he berated the visitor. Whereupon Russell, so far silent, suddenly sprang to Popper's defense, his "bushy white hair crowning his stance" as he "roared forth like a Sinaitic god."

In most of these accounts the poker is imprinted on the witness's mind. But only John Vinelott sees the crucial point—whether Popper makes what was probably an attempt at a joke to Wittgenstein's face—in Popper's way. Vinelott hears Popper utter his poker principle and observes that Wittgenstein is clearly annoyed at what he thinks is an unduly frivolous remark. Wittgenstein leaves the room abruptly, but there is no question of the door being slammed.

Up against these versions stands Karl Popper's testimony, a detailed narrative in which he sees how Wittgenstein uses the poker for emphasis, how he demands a statement of a moral principle, and how he, Popper, responds, "Not to threaten visiting lecturers with pokers." He sees Wittgenstein throw down the poker and storm out, slamming the door.

How does Professor Geach deal with these divergent accounts? Manifesting the depth of passion that the incident still incites, he declares simply that Popper lied. For Geach, the crucial issue is straightforward: whether Wittgenstein left the meeting after Popper cited the poker-threat principle, as Popper claimed. Geach is certain that he saw Wittgenstein leave before that.

For his part, Professor Watkins displayed some uncertainty about his version after being challenged in the *Times Literary Supplement*. Following further research, he wrote to say that he was prepared to reserve judgment on exactly when Wittgenstein left the meeting—"as a matter of detail." It was a risky concession. After all, in Popper's autobiographical account Wittgenstein's rage had probably been caused by Popper's joking—logically impossible if Wittgenstein left before the joke. In the event, as in cross-examination, to concede served only to move the advocate to still higher levels of scorn and further criticism of the witness. Disdainfully Geach rejoined, "If somebody falsely says 'John and Mary had a baby and then got married,' he would not be very well defended by a friend who said his memory might have slipped as to whether the birth or the marriage came first."

On crucial elements of the story—the sequence of events, the atmosphere, how the antagonists behaved—there are clear memories equally clearly in conflict. The poker is red-hot or it is cool. Wittgenstein gesticulates with it angrily or uses it as a baton, as an example, as a tool. He raises it, uses it for emphasis, shakes it or fidgets with it. He leaves after words with Russell or he leaves after Popper has uttered the poker principle. He leaves quietly or abruptly, slamming the door. Russell speaks in a high-pitched voice or he roars.

What really happened, and why?

3

Bewitchment

God has arrived. I met him on the 5:15 train.
—JOHN MAYNARD KEYNES

He cast a spell.
—FANIA PASCAL

ONE PROBLEM IN ATTEMPTING an evenhanded treatment of the two antagonists is what can only be described as Wittgenstein's capacity to bewitch, reaching out across the decades to demand attention.

Partly his spell is transmitted through the gleam in the eye and the lift in the spirits of his former students as they recollect him and the grip he still holds over them. Partly it comes through his riddling utterances, which lend themselves to an endless process of interpretation and reinterpretation. Partly it is in the complexity of his personality as it passed down to us through reminiscence and commentary—"An arresting combination of monk, mystic and mechanic," wrote the literary theorist Terry Eagleton, the author of a screenplay and novel about Wittgenstein.

The image of Wittgenstein as a religious figure, a seer, saint-like and suffering for mankind, runs through many accounts of him, whether fact or fiction. He told the economist John Maynard Keynes that he had given up philosophy to teach in a village school in Austria in the 1920s because the pain that teaching gave him overcame the pain of doing philosophy—as a hot-water bottle pressed against the cheek takes away the pain of a toothache. In the feline comment of the philosopher and social anthropologist Ernest Gellner, "Wittgenstein's place was achieved by his suffering." In Jewish terms, he could be seen as a traditional wilderness-wandering *tsaddik*, a holy man. He is portrayed in one novel as "the desert mystic, subsisting on bread, rainwater and silence."

But to leave the characterizations there would be to mislead. Above all, Wittgenstein comes down to us as dynamic and powerful. Those who knew him—both friends and foes—describe him in language that is without moderation. And the invocation of Wittgenstein in a stream of literary and artistic works outside philosophical publications is a striking confirmation of the hold he exercises long after his death. In trying to understand this fascination, perhaps the secret is to see Wittgenstein as a literary figure who fits as easily into a discourse on authors—such as Proust, Kafka, Eliot, Beckett—as into a study of philosophers.

In her book on twentieth-century poetic language, *Wittgenstein's Ladder*, the American critic Marjorie Perloff instances

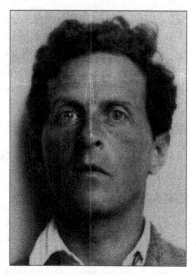

Ludwig Wittgenstein, taken at around the period of the poker meeting. His students found him "incandescent with intellectual passion."

eight novels and plays, twelve books of poetry, and some six performance pieces and experimental artworks that are directly about or influenced by Wittgenstein. And, chronicling the paradoxes of Wittgenstein's life, she comments, "It is, no doubt, one that lends itself to dramatic and fictional representation, to the making of myths. For Wittgenstein comes to us as the ultimate modernist outsider, the changeling who never stops reinventing himself." To put it another way, Wittgenstein can be what we want him to be.

He might also be unique among philosophers in having become part of hard-pressed journalists' shorthand, with his name standing in for "charismatic genius." A 1990s style-setter was described as "a restaurateur with the mesmeric hold of a Wittgenstein." "You don't have to be Wittgenstein to understand . . ."

offers an alternative to "You don't have to be a rocket scientist . . . "; while "He's no Wittgenstein" puts someone in his intellectual place. The architect Sir Colin St. John Wilson, whose designs have been heavily inspired by Wittgenstein, though he never met him, says, "He was obviously a magician and had qualities of magic in his relations with people."

Wittgenstein's imprint on those he taught is reflected in a story told by Peter Gray-Lucas, whom we encountered in H3. Gray-Lucas was no admirer of Wittgenstein, regarding him as a "charlatan." Even so he found his personality compelling:

> He was an absolutely marvellous mimic. He missed his vocation: he should have been a stand-up comedian. In his funny Austrian he could do all sorts of mimicry of accents, styles, ways of talking. He was always talking about the different tones of voice in which you could say things, and it was absolutely gripping. I remember one evening he got up from his chair, talking in this funny voice, and said something like, "What do we say if I walk through this wall?" And I remember realizing that my knuckles were going white gripping my armchair. And I really thought that he was going to go through the wall and that the roof was going to fall in. That must have been part of his spell: that he could conjure up almost anything.

Another part of his spell is that he appeared able to achieve originality and excellence wherever he took an interest. As a young engineering student in 1910 he patented a novel aircraft engine that anticipated the jet engine and was reinvented and tested successfully in 1943. In the First World War he became a much-decorated fighting soldier. Between the wars he compiled an innovative dictionary for primary-school children and played a major part in the design of a much-praised modernist house. In the Second World War, working as a laboratory assistant in a medical team researching into wound shock, he devised new apparatus for measuring changes in respiration brought on by changes in blood pressure. Wherever Wittgenstein went he left his creative mark.

Karl Popper's presence is not to be found haunting stage plays and poetry. In truth, such a thing is hard to imagine: he could scarcely be a greater contrast to Wittgenstein, presenting a picture of sheer human ordinariness, with an undeviating academic and married life. As for their impact on others, while Wittgenstein would at once dominate any room he entered, Popper could pass almost unnoticed, as his friend and champion the philosopher, politician, and broadcaster Bryan Magee recollects from his first sight of him at a meeting:

> The speaker and chairman entered side by side. At that moment I realized I did not know which of the two was Popper. . . . However, since one was a solid, self-confident figure and the other small and unimpressive, it looked as though the former must be Popper. Needless to say, it was the latter, the little man with no presence. However, he lacked presence only for so long as he was not speaking—though even then what compelled attention was not his manner but the content of what he said.

This disparity between Popper's, at first glance, diffident demeanor and his zeal on stage and in debate struck his successor at the London School of Economics. Reflecting on the poker incident, John Watkins described Popper as having "a bit of a cat and lion, oscillatory tendency. First there's a little man, looking perhaps rather frightened, or rather apprehensive and insecure.

And in no time at all he's blowing himself into this big challenge." That preliminary diffidence might have been something to do with Popper's self-image. Not only was he short, he was of curious build—small legs and big chest. Moreover, "Physically he had these great long ears. For a long time he was very worried and had an inferiority complex about his looks." In later life he apparently enlarged his ears still more by tugging the lobes in order to hear better. Some maintain that his wife, Hennie, made him feel inadequate by not showing the affection he craved.

A final point of comparison concerns the immediacy of the two antagonists' work. Wittgenstein's pithy exclamations questioning our thoughts—like the words of an oracle—continue to compel attention. Popper's great contributions to politics and to our understanding of history and of scientific method—written in plain English prose—have to some extent been overtaken by time and eroded by critics. The fall of the Berlin Wall and the implosion of Communist regimes justify Popper's theoretical dismemberment of totalitarian government and his advocacy of the open society. But the very success of his endeavor leaves him as a figure of past greatness rather than present influence.

Each has much to say to us today, but two examples from the press demonstrate their present/past relevance. The last issue of the *Spectator* for the twentieth century contained no fewer than three contemporary cultural references to Wittgenstein, including one identifying his later philosophy as the inspiration for Michael Frayn's best-selling comic novel *Headlong*. At the same time a *Financial Times* article looking back at the past century depended on Popper for an analysis of the link between the century's horrors and advances. The author, Martin Wolf, found the link in Popper's vision of the confrontation between freedom and the fear of it and between the open and closed society. The positive aspects of the twentieth century—the progress of science, the flow of innovation, the spread of democracy, decolonization—were the products of the open society. The negative came from the reaction against it—the willingness to kill millions of people because they failed to fit into an uncompromising vision of an ideal world.

So the abiding spell of Wittgenstein should not obscure the fact that Professor Sir Karl Popper CH, FRS, FBA was widely saluted in his lifetime as one of the world's most original thinkers.

Karl Popper in 1946. Newly arrived in Britain from teaching in New Zealand, he was winning a growing band of admirers for his book demolishing totalitarianism, The Open Society and Its Enemies.

Cambridge University
Moral Science Club.

FOUNDED 1874.

"Dissere aut Discede".

RULE II.

The main object of the Club is the periodical discussion of Philosophical Subjects, such discussions to be *introduced* by the reading of ... papers ... and it is therefore desirable that the papers should be as short as possible.

MICHAELMAS TERM, 1946.

President:
Prof. C. D. BROAD

Chairman:
Prof. L. WITTGENSTEIN

Hon. Sec.:
W. HIJAB.

759.46   Metcalfe's, Trinity Street

The Moral Science Club term card. The club offered abstruse philosophy papers delivered by eminent thinkers to serious students.

4

Disciples

[Popper] was the Socrates of our time.
— ARNE PETERSEN

Reading the Socratic dialogues one has the feeling: what a frightful waste of time!
— WITTGENSTEIN

POKER SHAKING, DOOR SLAMMING — what sort of academic forum was the Moral Science Club?

The minutes from 1878, preserved in the Cambridge University Library, show that it was (and still is) a club well used to arcane debate led by eminent thinkers. The week after the poker incident the speaker was an Oxford don, J. L. Austin, the leader of the ordinary-language school of philosophy that found value in exploring the nuances of everyday speech. He devoted his talk to a peculiar phenomenon in language: first-person-singular present-indicative verbs whose very utterance—"I name this ship *Queen Elizabeth*," "I declare this meeting open," "I do" (in a wedding ceremony)—constitutes an act. Other papers delivered around this time examined the possibility of hallucinations, the gap between appearance and reality, and the idea of certainty. Earlier in the year A. J. Ayer had spoken on the nature of causation.

Anyone who attended MSC evenings demonstrated a seriousness about philosophy that went beyond the call of duty for the average undergraduate. Then, as now, the club competed for attention with a plethora of other attractions. There was watery beer to be drunk (though there was little to be found). There were debates to be joined, music to be played, magazines to be edited, politics to be argued over. There was the call of the stage, the river, and the sports field. There were even essays to be written. After a day of lectures and tutorials, the prospect of two hours on first-person-singular present-indicative verbs was enticing only to the most devoted and conscientious. With such an audience, those who presented papers could expect rigorous challenge.

But in the 1930s and 1940s that was not the sole reason why meetings were only for the strong-minded: according to some accounts, they were permeated by a tribalism normally associated with the football terrace rather than an esoteric academic club. Passionately expressed allegiance to Wittgenstein, it was said, bubbled up through every discussion. The philosopher Gilbert Ryle noted that on his occasional visits to the club "veneration for Wittgenstein was so incontinent that my mentions of any other philosopher were greeted with jeers."

Some MSC regulars dispute this. While strong opinions might

have been expressed, says Sir John Vinelott, debate was always mannerly. Georg Kreisel agrees: strong yet civil. However, delivering a paper to the club could be destructive for the speaker even when it was not punctuated with jeers. On 12 June 1940, just as German tanks stormed through the French defenses and opened the way to Paris and the Channel, Isaiah Berlin ventured from All Souls College, Oxford, to the MSC. His biographer Michael Ignatieff describes the scene:

> All the Cambridge philosophers turned out—Braithwaite, Broad, Ewing, Moore, Wisdom and a sixth figure, small and handsome of feature, who appeared surrounded with acolytes in tweed jackets and white open-necked shirts identical to his own. This was Ludwig Wittgenstein. Berlin delivered his paper on the problem of how could one have knowledge of others' inner mental states. After a few initial questions, Wittgenstein became impatient and took over. Berlin remembered him saying, "No, no, that is not the way to go about it. Let me. Don't let's talk philosophy. Let's talk business with each other. Ordinary business . . . ."

After an hour, Wittgenstein rose to his feet, his acolytes rose with him, and he leaned over the table and shook Isaiah's hand. "Very interesting discussion. Thank you." With that he walked out. Their encounter marked the symbolic end, if not the actual end, of Isaiah's active philosophical career.

That there was a band of vocal supporters for Wittgenstein seems unsurprising. Here was a magnetic teacher. Like the radical English don F. R. Leavis, his Cambridge contemporary, who spent many hours in the 1930s walking and talking with him, Wittgenstein attracted disciples rather than students. Just as with Leavis, they tended to imitate his mannerisms. Wittgenstein's successor as Professor of Philosophy, Georg Henrik von Wright, recorded, "Wittgenstein himself thought that his influence as a teacher was, on the whole, harmful to the development of independent minds in his disciples. I am afraid that he was right. To learn from Wittgenstein without coming to adopt his forms of expression and catchwords and even to imitate his tone of voice, his mien and gestures was almost impossible."

One of Wittgenstein's students, Norman Malcolm, later Professor of Philosophy at Cornell University—a friend whom Wittgenstein believed to be "a serious and decent man"—reached the same conclusion: "Few of us could keep from acquiring imitations of his mannerisms, gestures, intonation, exclamations." These included the putting of his hand over his forehead, the cry of "*Ja!*" when there was something he emphatically approved of, and the intensity with which he wrinkled his brow. He would push his hands out, flat against each other, fingers outstretched, toward someone he agreed with; disagreement was signaled with a sharp downward movement.

A notable anecdote about this mimicry relates to Wittgenstein's influence on Malcolm himself. In 1949, when Wittgenstein visited Malcolm at Cornell and sat in on one of his seminars, a student asked who the old guy was at the back—"impersonating Malcolm." The imprint Wittgenstein made ran deep. A decade after his death, Fania Pascal, who taught him Russian in the 1930s and became a friend, recognized his characteristics in a chance new acquaintance—and, what is more, a nonphilosopher.

Then there were the shirts, the top button carefully undone. Sir John Vinelott remembers the acolytes as being scruffier versions of Wittgenstein, who might have dressed casually but was neatness itself: "I thought when I first met him that he looked like a retired army officer. He had an open-necked shirt, a tweed jacket, grey flannel trousers, well-polished brogues. But none of it was the least bit scruffy. It was all fastidiously kept, and his turnout was exceptional."

How religiously the inner circle of the disciples followed the master has a comical air to it: sleeping in narrow beds, wearing sneakers, carrying vegetables in string bags to let them breathe, and putting celery in water when serving it for dinner. But this was not just a matter of amusing idiosyncrasies being aped: his students were also liable to change their attitude toward life in general, adopting Wittgenstein's "proud and even contemptuous austerity," learning to scorn what they had previously enjoyed as harmless luxuries as "utterly trivial and unworthy of attachment." In fact they could be more Wittgensteinian than Wittgenstein, for the master himself was not as ascetic as is often presented. Take the story that he never dined at high table in Trinity, as it would have meant putting on a tie. A student of Wittgenstein's in the late 1930s, Theodore Redpath, who became a Cambridge don, recounts how Wittgenstein borrowed his tails, white waistcoat, tie and stiff shirt to attend the Trinity Fellowship Admission Dinner in October 1939. He was going "as a *Professorial* Fellow," he told Redpath, "with typical mock-pride" (Redpath's emphasis).

ALTHOUGH NOTORIOUS for taking over the meetings, Wittgenstein did on occasion listen at the Moral Science Club—and learn. In 1944 G. E. Moore delivered a paper in which he raised a puzzle that Wittgenstein considered of fundamental importance, naming it "Moore's Paradox" and devoting an evening to it at a subsequent session on 25 October 1945, one year to the day before the Popper clash. Moore then replied to Wittgenstein's reply, entitling his talk "P but I do not believe P."

"Moore's Paradox" addressed such propositions as "Smith left the room but I don't believe it" and "There is a fire in this room and I don't believe there is." Moore thought these absurd, because they were psychologically impossible. But Wittgenstein's excitement about them was that they were logically impermissible even though they were not of the form "Smith left the room and Smith stayed in the room." They defied the logic of our language: nobody would utter such a sentence. In other words, thought Wittgenstein, they showed that propositions could be disqualified from use even when they were not strictly contradictory—that is, even though they were not of the form "P and not-P." This demonstrated to Wittgenstein that what was impermissible in language was much more subtle than he had previously believed—that there was more to commonsense logic than the formal logic practiced by logicians.

Moore, like the other luminaries of Cambridge philosophy, used the MSC as a forum in which to test fledgling ideas. Depending on how they fared, they could later be tinkered with or abandoned altogether. As for Wittgenstein, if the topic under discussion caught his interest he would become utterly engrossed, oblivious to his surroundings. On one occasion, when he was walking home with Michael Wolff after an MSC meeting, two speeding U.S. Army lorries passed close enough to make Wolff's gown flutter. "Those lorries go too fast," he grumbled. Totally unconscious of the near miss, Wittgenstein assumed that Wolff's comment was a metaphor about the MSC paper and replied, "I can't see what that has to do with the question."

To make MSC meetings as productive as possible, Wittgenstein had firm views on how they should be run. In 1912, a year after his arrival at the university to study under Russell, he had imposed his will on the club, pushing through a plan for there to be a chairman to "guide" discussion. G. E. Moore was elected to fulfill this role (a post he held for thirty-two years). Wittgenstein's aim had been to stamp out posturing and empty verbiage, and his preference throughout his Cambridge career was for papers to be as short as possible. He began as he demanded others should continue. Toward the end of 1912 he read a paper in his own rooms on "What is Philosophy?" "The paper," say the club's minutes, "lasted only about four minutes, thus cutting the previous record established by Mr. Tye by nearly two minutes. Philosophy was defined as all those primitive propositions which are assumed as true without proof by the various sciences. This definition was discussed, but there was no general disposition to adopt it."

Over the next thirty-five years, Wittgenstein's relationship with the Moral Science Club was, like all his relationships, tempestuous and unpredictable. In the early 1930s he stopped attending meetings, after complaints that no one else could get a word in edgeways. When he discovered that Russell was to address the MSC in 1935, he wrote to his former mentor to explain his predicament. It would be natural for him to attend, he says:

But:—(a) I gave up coming to the Mor[al] Sc[ience] Cl[ub] 4 years ago; people then more or less objected to me for talking too much in their discussion. (b) At the meeting there will be Broad, who, I believe, objects most strongly to me. On the other hand (c), if I am to discuss at all it will—in all likelihood—be the only natural thing for me to say a good deal, i.e., to speak for a considerable time. (d) Even if I speak a good deal I shall probably find that it's hopeless to explain things in such a meeting.

After Moore stepped down as chairman in 1944 for health reasons, Wittgenstein became his successor. During the next two years he would be replaced, only to be reelected. By then, however, his attitude to the meetings had changed. Norman Malcolm said Wittgenstein found the atmosphere "extremely disagreeable."

He went only out of a sense of duty, thinking that he ought to do what he could to help make the discussions as decent as possible. After the paper was read, Wittgenstein was invariably the first to speak, and he completely dominated the discussion as long as he was present. He believed that it was not good for the club that he should always play such a prominent role there, but on the other hand it was quite impossible for him not to participate in the discussions with his characteristic force. His solution was to leave the meetings at the end of an hour and a half or two hours. The result was that the discussion was exciting and important while Wittgenstein was present, but trivial, flat, and anti-climatic after he left.

The club needed to find an answer to Wittgenstein's domination of its meetings. At various times during his presence in Cambridge the MSC adopted, with his support, a system of starring certain meetings, at which members of the faculty were "not expected to appear." Although the star was in theory designed to exclude all the dons, in practice everybody understood that it was aimed at one person only. Wittgenstein certainly intimidated the students, and the dons complained that his habit of interrupting speakers was also very discourteous to visiting lecturers. Even when a star appeared by a paper on the term card, the other members of the faculty found ways of circumventing the regulations. They would on occasion show up as guests of one or other of the students.

Popper's paper, however, was not starred — nor were any papers that term. Nevertheless, other rules were in force, having been laid down after Wittgenstein became chairman. Then he had dictated the format for the invitation to guest speakers, specifying "short papers, or a few opening remarks, stating some philosophical puzzle." This wording was designed to fit both his distrust of formal lectures and his view of the proper bounds of philosophical discourse: there were no real problems of philosophy, only linguistic puzzles. Wasfi Hajib had followed Wittgenstein's wording to the letter in his invitation to Popper.

As the audience of undergraduates and dons packed into H3 on that October evening were to discover, Dr. Popper had given that invitation a close reading.

5

The Third Man

Then Russell appeared — to inform me of some alterations he is making in the hours of his lectures — and he and Wittgenstein got talking — the latter explaining one of his latest discoveries in the Fundamentals of Logic — a discovery which, I gather, only occurred to him this morning, and which appears to be quite important and was very interesting. Russell acquiesced in what he said without a murmur.

— DAVID PINSENT

FROM HIS HIGH-BACKED ROCKING CHAIR in front of the fireplace, an elder statesman of philosophy calmly watched the conflict between Popper and Wittgenstein. In this story he is the Third Man, the Cambridge connection between the two Viennese.

At seventy-four, he was unquestionably much better known to the public than were the two antagonists. The shock of white hair, the refined birdlike features and the customary pipe would have made Lord Russell — Bertrand Russell — instantly recognizable to millions beyond that room who had seen him in newsreels and newspaper photographs, whereas even fellow philosophers might have been hard-pressed to identify Popper and Wittgenstein. He was certainly as eminent as those two. Indeed, it can be argued that Russell was the true audience that day for both Popper and Wittgenstein. Russell was scarcely acquainted with Popper, though he had given him a helping hand; Wittgenstein, whom he had befriended many years earlier, he had come to know intimately. Both men owed him a debt. Popper's debt was small, though he felt immense gratitude. Wittgenstein's debt was immense; but by 1946 he felt for Russell only barely concealed contempt.

While Popper and Wittgenstein were both exiles in Britain, betraying their Austrian origins whenever they opened their mouths, Russell was the exemplar of Englishness. The grandson of Lord John Russell, a nineteenth-century Liberal prime minister, Bertrand Arthur William was born in 1872 into the upper tier of Victorian social and political life. At the home where he spent his childhood he was accustomed to leading politicians dropping in. Once, after the ladies had retired from the dinner table, he was left alone to entertain the grand old man of British politics, William Ewart Gladstone. Gladstone spoke to the child only once: "This is very good port they have given me, but why have they given it to me in a claret glass?" With such a background, it came naturally to Russell to consort with the great and the good. When he had a request to make, a policy to push, or a cause to fight, he simply wrote personally addressed letters to national leaders. Russell was not socially intimidated or intellectually over-awed by anyone.

In his early thirties Russell had made his academic reputation with his pioneering work in logic and mathematics. He also has a strong claim to be considered the father of analytic philosophy, which has come to dominate Anglo-American thought. For that reason alone, his position in the pantheon of philosophy is as-sured. He may now be quoted rarely and acknowledged less, but most mainstream philosophers of today are operating within the framework that he established.

His subsequent fame extended far beyond academia. It was based on his political activities and his popular writings, which covered a bewildering array of subjects from marriage and reli-gion to education, power, and happiness. His output throughout his life was prolific: he would turn out a book or two a year—some of them weighty tomes, some popular tracts. His effortless style— funny, mischievous, polemical, and always crystal clear—won him an international following and, in 1950, the Nobel Prize for Literature.

His books could land him in trouble. Only two years before the poker clash Russell had returned to Cambridge from a miserable spell in the United States, during which he had been blocked from becoming a professor at New York's City University. A Catholic mother, supported by the religious hierarchy, claimed his teaching could do untold damage to her daughter. The woman's lawyer quoted from Russell's work, summing it up in overblown courtroom rhetoric as "lecherous, libidinous, lustful, venerous, erotomaniac, aphrodisiac, irreverent, narrow-minded, untruthful and bereft of moral fiber." It would have been funny had it not cost Russell a job. But when Russell shortly afterward published *An Inquiry into Meaning and Truth*, listed on the cover was an impressive catalog of his philosophical qualifications and then, at the bottom, a wry addition: "Judicially pronounced unworthy to be Professor of Philosophy at the College of the City of New York."

Russell was not averse to a scrap: naturally outspoken, with a mind that was always faster, sharper, and more nimble than his opponents, he lived his life in pursuit of contentious causes. During the First World War he was sent to jail for writing an article that suggested that the American troops deployed in Britain would subsequently be used as a strike-breaking force if workers sought an end to the war through industrial unrest. He then exploited his social connections to ensure that he served his term in the most relaxed of prison conditions, with a cell to himself, food sent in from outside, and unlimited books—unlike the conscientious objectors whom he had encouraged and whose suffering in jail frightened him when his turn came. He used the calm of his confinement to return to philosophical study.

Later, well into his eighties, he was sentenced to prison again—this time for pursuing a policy of civil disobedience as part of his dogged campaign against nuclear weapons (though not long before the 1946 meeting he had argued for their use against the Soviet Union, so worried was he by the development of the Russian nuclear program). He was the first president of the Campaign for Nuclear Disarmament, and helped found the Pugwash Conferences, at which distinguished intellectuals discussed how to guarantee world peace. In his old age, his high profile and unrestrained hostility to the Vietnam War could still arouse a mixture of anxiety and rage in the political establishment.

Russell managed all this while standing for election three times (once on a platform for women's suffrage), globe-trotting, broadcasting, lecturing, opening and running a school, receiving a chestful of honors, marrying four times, having children, and conducting several vigorous affairs which (to his delight) scandalized polite society. He also wrote literally tens of thousands of letters, many of which have found an afterlife in the archives. He would reply to almost all the members of the public who corresponded with him, whether they had praised him or, as was not uncommon, chastised him. A typical note came from a Mrs. Bush, who had just read his autobiography. "Thank you," she wrote. "I have already thanked God." To which Russell replied, "I am pleased that you liked my autobiography, but troubled that you thanked God for it, because that suggests He has infringed my copyright." (He also replied to a fourteen-year-old schoolboy—one of the authors—who asked for his help in understanding how space could have a boundary. Russell directed him to non-Euclidian geometry.)

Given his eminence in philosophy and the range of his activities, it should be no surprise that Russell knew both Wittgenstein and Popper. But his relevance to the happening in H3 is that he had actively helped both of them and that they would conceivably not have faced each other there without their contact with him. In Wittgenstein's case, it is no exaggeration to say that the course of his life was changed by his contact with Russell.

By 1911 the twenty-two-year-old Ludwig had become preoccupied with the philosophy of mathematics. His father had wanted him to have a technical education, so he had already spent two years in Berlin and three years in Manchester studying aeronautics, building experimental kites, and eventually designing an airplane engine. Now he felt compelled to turn to philosophy, and, following conversations with British and German mathematicians—including Gottlob Frege—he sought out the internationally renowned logician, the Honorable Bertrand Russell of Trinity College, Cambridge.

Some eight weeks later, having passed the autumn term as a guest student at Trinity, Wittgenstein requested the answer to a simple question: was he utterly hopeless at philosophy? Russell did not know what to think. Wittgenstein returned to Vienna to write something for Russell to look at. The result of his endeavor, declared Russell, was "very good, much better than my English pupils do. I shall certainly encourage him. Perhaps he will do great things."

By the summer of 1912, within six months of Wittgenstein taking up his full-time place, Russell had come to believe that he had discovered his intellectual heir. Wittgenstein was, he thought, "perhaps the most perfect example I have ever known of genius as traditionally conceived, passionate, profound, intense and dominating." He later repeated this view to an American confidante, Lucy Donnelly: "His avalanches make mine seem mere snowballs. . . . He says every morning he begins his work with hope & every evening he ends in despair—he has just the sort of rage when he can't understand things that I have."

Soon their teacher-student roles were reversed; for the first time in his life, Russell felt intellectually overpowered. In 1916, in a letter to his lover, the socialite Lady Ottoline Morrell, he mentioned an incident three years earlier, when Wittgenstein had been severely critical of some work Russell had been doing in the field of epistemology. Although Russell could not fully comprehend Wittgenstein's inarticulate comments, they were sufficient to convince him that he himself was wrong:

> His criticism, tho' I don't think he realized it at the time, was an event of first-rate importance in my life, and affected everything I have done since. I saw he was right, and I saw that I could not hope ever again to do fundamental work in philosophy. . . . Wittgenstein persuaded me that what wanted doing in logic was too difficult for me.

Not long after he had met Wittgenstein, Russell said in a letter to Lady Ottoline, "I love him & feel he will solve the problems I am too old to solve." And, after a year, Russell told Ludwig's eldest sister, Hermine, who was in Cambridge visiting the baby of the family, "We expect the next big step in philosophy to be taken by your brother."

Their early relationship was one of mutual respect and affection. Russell was an emotional sheet anchor for Wittgenstein, who would often go to Russell's rooms and pace up and down in silence. "Are you thinking of logic or your sins?" Russell once asked. "Both," came the reply. Sometimes Wittgenstein's mood was so savage that Russell feared he would break all the furniture in his room.

His anxiety that Wittgenstein would suffer a breakdown—even kill himself—was well founded: Wittgenstein confessed his suicidal feelings to his friend David Pinsent, a mathematics student at Trinity. After returning to Cambridge from a visit to Norway in 1913, Wittgenstein informed Russell that, as soon as he could, he would head straight back to the fjords and would then live entirely alone until he had answered all the questions of logic. Russell tried logic to dissuade him: he pointed out it would be dark. Wittgenstein had an answer: he hated daylight.

> I said it would be lonely, & he said he prostituted his mind talking to intelligent people. I said he was mad & he said God preserve him from sanity (God certainly will). Now Wittgenstein, during Aug. and Sep. had done work on logic, still rather in the rough, but as good, in my opinion, as any work that ever has been done in logic by anyone. But his artistic conscience prevents him from writing anything until he has got it perfect, & I am persuaded he will commit suicide in February.

The feeling of being on a precipice above insanity was one that Russell fully understood. Madness was in his family, and he often felt that he himself was in danger of tipping over the edge. Lady Ottoline thoughtfully dispatched a recipe for cocoa that she believed would calm the Austrian's strained nerves and ease his depression. Russell thanked her, but it is not clear whether Wittgenstein tried it. If so, it did not fulfill Lady Ottoline's expectations.

Although Wittgenstein was never the easiest of companions, he had the effect on Russell of reenergizing his intellectual batteries. "Wittgenstein makes me feel it is worth while that I should exist, because no one else could understand him or make the world understand him." And, equally importantly, Russell now believed that he had at last found somebody capable of continuing his work, and he declared himself content to bequeath the future of logic to the younger man.

The enormously high regard in which Russell held his former

student would prove vital to Wittgenstein. In the only book of philosophy Wittgenstein published in his lifetime, the *Tractatus Logico-Philosophicus*, composed in the trenches of the Great War, the author modestly concluded that he had solved all the essential problems of philosophy. As he was not yet thirty, this was no mean boast, but it was not one that convinced publishers, and the *Tractatus* would not have seen the light of day without Russell's practical assistance. Although the individual sentences of the *Tractatus* have a deceptive simplicity, the work as a whole is opaque to the lay reader, and not much more transparent to the specialist. After the war a print run was agreed with a German publisher, Wilhelm Ostwald, but only on condition that Russell would write an introduction, explaining why the book was important. This Russell did, though including some reservations.

Contact between Russell and Wittgenstein had been reestablished after Wittgenstein's capture, along with thousands of other Austrian soldiers, in Italy. Wittgenstein spent part of 1918–19 in a prisoner-of-war camp, but, once he got word of his whereabouts out to Russell, the latter, with assistance from Keynes, arranged for him to have correspondence privileges. This enabled Wittgenstein to send Russell his manuscript. Russell and Wittgenstein then met after Wittgenstein's release, going through every one of the book's propositions. Despite that exercise, when the author finally read Russell's introduction he was irate—his former teacher, he felt, had completely missed the point. Nevertheless, Russell's imprimatur had made all the difference: the *Tractatus* was published in German in 1921, and an English translation by C. K. Ogden followed in 1922.

Wittgenstein by then felt mentally drained. He had spent seven years on the ideas that culminated in the *Tractatus*, and believed that with this work his contribution to philosophy was complete—there was, as he put it, no more to be squeezed from the lemon. It was only when he began to think about philosophy afresh in the years between 1927 and 1929, stimulated by conversations with the founder of the Vienna Circle of logical-positivist philosophers, Moritz Schlick, that he decided to return to Cambridge. Russell, along with Keynes, was instrumental in bringing him back.

Even during the six years from 1920 more or less lost to philosophy, when Wittgenstein was variously a schoolteacher, a monastery gardener, and an architect, there were contacts with Russell and some of the Cambridge set. The precociously brilliant mathematician Frank Ramsey visited him in the mountain village of Trattenbach in lower Austria—at the time Ramsey was only nineteen—and sent back news to Russell of Wittgenstein's state of mind and ascetic lifestyle. Russell and Wittgenstein also corresponded directly. One exchange shows Russell's skepticism of Wittgenstein's insistent claim that the people of Trattenbach, where he was teaching, were uniquely despicable.

When Wittgenstein returned to Cambridge in 1929, Russell was, initially at least, again of enormous help. The *Tractatus* was submitted as a doctoral thesis. Russell and G. E. Moore, whom Wittgenstein knew well from his first spell at the university, were Wittgenstein's examiners in a process that could most generously be described as a sham. At the viva stage, when Wittgenstein had to be questioned on the arguments of his thesis, the three acquaintances sat for some time chatting, before Russell turned to Moore and said, "Go on, you've got to ask him some questions— you're the professor." A desultory discussion ensued, at the end of which Wittgenstein stood up, slapped his examiners on the shoulder, and said, "Don't worry. I know you'll never understand it." When Dr. Wittgenstein had completed his first academic year on a one-off study grant from Trinity, Russell was asked to produce a report on his protégé's work; the outcome was a research fellowship.

All this Russell did gladly for Wittgenstein—he was his mentor, sponsor, therapist, and supplier of recipes for curative hot drinks. Yet by 1946 the relationship had long since soured. The involvement and passion of the late-night discussions of 1911 to 1913 had been replaced by an icy distance, brought on by what Wittgenstein perceived as their irreconcilable personalities.

Wittgenstein found Russell's approach to philosophy too mechanistic, and his approach to people too emotionally rootless. But if there was one particular facet of Russell's personality that was intolerable to him, it was a certain glibness. Wittgenstein was incapable of being halfhearted about any activity in which he engaged. Russell, though he was a man of great principle and—unlike Wittgenstein—prepared to campaign for his values in public life, was not governed every second of every day by an unyielding personal morality. He was willing to compromise, to utter a small lie here, a small exaggeration there, to flatter if necessary, to soothe if required. These were means—minor transgressions—which could be justified by the ends he sought.

Symptomatic of this, thought Wittgenstein, were the popular moneymaking books which Russell churned out and which Wittgenstein loathed. In particular, he was offended by Russell's militant atheism and appalled by his freethinking homilies on marriage and sex. Of the latter, he said, "If a person tells me he has been to the worst of places I have no right to judge him, but if he tells me it was his superior wisdom that enabled him to go there, then I know that he is a fraud."

There was indeed some irony in Russell's sermonizing on relationships, given his lack of emotional insight and his fraught bonds with his family, who accused him of coldness, callousness, and cruelty toward them. Having decided during a bicycle ride that he no longer loved his first wife, Alys, he broke the news to her immediately on returning home. Though they divorced, she never fell out of love with him. His granddaughter has claimed that he slept with his daughter-in-law, breaking up the marriage of

Bertrand Russell in the 1940s. To Popper he was supreme as thinker and writer; to Wittgenstein, "still amazingly quick, but glib and superficial."

his son, John. He has been charged with driving John to madness, and of causing two of his wives to attempt suicide.

English manners being foreign to Wittgenstein, he expressed many of his forthright views about Russell bluntly to his face. These included his low opinion of all Russell's philosophical work since the First World War. A letter written from his prisoner-of-war camp in 1919 gives a flavor. Wittgenstein had just read a copy of Russell's latest book, *Introduction to Mathematical Philosophy*; the *Tractatus* was still in limbo. "It's galling," he wrote, "to have to lug the completed work round in captivity and to see how nonsense has a clear field outside."

WHATEVER WITTGENSTEIN'S ATTITUDE, by 1946 Russell was a player on the world stage: a totemic figure, a popular sage, his lectures and writings greedily consumed by a large public following. A year earlier, about the only happy outcome of his Second World War years in America had been the publication of his sweeping 900-page survey, *A History of Western Philosophy*. Albert Einstein wrote of it, "I regard it as fortunate that so arid and brutal a generation can claim this wise, honorable, bold and humorous man." It was a surprise best-seller, freeing Russell from financial worries. A letter "from the inner sanctum" of his American publisher, Simon & Schuster, dated 30 September 1946, reveals that by then it had already sold nearly 40,000 copies.

But, despite his public prominence, Russell found himself eclipsed in the much narrower academic circles where he still had a role he valued. Wittgenstein's ideas had moved on, and the preeminence of his new school had left Russell's philosophic work on the sidelines. As the elder statesman of philosophy put it, "It is not an altogether pleasant experience to find oneself regarded as antiquated after having been, for a time, in the fashion." Quite how much he understood of the later Wittgenstein's work is open to question. Stephen Toulmin overheard Russell asking Richard Braithwaite in 1946 what Wittgenstein had been doing since the *Tractatus*.

In vogue or not, Russell remained a draw to students. He may have seemed a relic, but he was a great relic, the Acropolis of a bygone philosophical age. His teaching at the time sprang from ideas that were to evolve into his book *Human Knowledge: Its Scope and Limits*. Academic critics gave it a mixed reception. Nevertheless, his classes were so packed that a second room had to be opened up and loudspeakers installed. In the Moral Sciences Faculty, where wit was in short supply, Russell's lectures were a tonic, spiced with jokes and seasoned with anecdote. Such was his love of talking to undergraduates that groups would assemble at the great gate of Trinity and set off across the normally forbidden lawns, listening avidly as he held forth.

Like most people who knew Wittgenstein, Russell had for a time been under his spell, blinded by his force. But in retrospect he took a bleaker view, describing him as very singular and saying, "I doubt whether his disciples knew what manner of man he was." And he accused Wittgenstein of debasing philosophy in his later years, and of "treachery" toward his own greatness. In his obituary of Wittgenstein, published in *Mind*, Russell wrote, "Getting to know Wittgenstein was one of the most exciting adventures of my life." But the article ends just after the publication of the *Tractatus*. On the remaining three decades of their acquaintance, and Wittgenstein's later work, he preferred to remain mute.

For his part, by 1946 Wittgenstein no longer believed Russell capable of first-rate work. Following a meeting of the Moral Science Club a few weeks after Popper's appearance, Wittgenstein sent Moore a letter. "Unfortunately (I believe) Russell was there and most disagreeable. Glib and superficial, though, as always, astonishingly quick." It was probably the last time the two saw each other: Wittgenstein told the American philosopher O. K. Bouwsma that they "passed but did not speak."

But a grudging esteem, rooted in early memories of Cambridge and shared logical endeavor, survived until the end. In 1937 Wittgenstein recorded in his notebook, "In the course of our conversations Russell would often exclaim: 'Logic's hell!'—And this *perfectly* expresses the feeling we had when we were thinking about the problems of logic; that is to say, their immense difficulty, their hard and *slippery* texture."

It was said that Wittgenstein remained more deferential toward Russell than toward any other person. Although he himself was permitted to upbraid Russell in public and to criticize him behind his back, those of his followers who followed suit would receive a stern rebuke.

IF RUSSELL—initially silent in H3 and well past the peak of his career—saw Wittgenstein in the light of their deep and complex relationship, when he looked at Popper he faced someone he scarcely knew but who was intent on a deep relationship with him.

The contacts between Russell and Popper had hitherto been perfunctory but cordial. This was unsurprising. For a start, there was the wide age gap between them—thirty years—with the result that professional jealousy was never an issue. Russell's first book, on German social democracy, had been published six years before Popper was born.

Russell also facilitated Popper's career, though in a relatively marginal capacity. The two men had met briefly at a philosophy conference in France in 1935, and then again in 1936 at a meeting of the Aristotelian Society in England. After this, Russell provided Popper with a testimonial when the younger man was desperately searching for a full-time job in order to escape Vienna. The vague and formulaic phrasing of the reference suggests that Russell felt barely acquainted with Popper's work: "Dr. Karl Popper is a man of great ability, whom any university would be fortunate in having on its staff." It went on, "I learn that he is a candidate for a post at Canterbury University College, Christchurch, New Zealand, and I have no hesitation in warmly recommending him." It has the feel of an off-the-shelf note that someone habituated to being used in this way might dash off without thinking.

Russell had received complimentary copies of both *The Logic of Scientific Discovery* and *The Open Society and Its Enemies*. His familiarity with the first is doubtful, if only because the pages of the copy in his library remained virtually uncut. And when Popper asked him to recommend *The Open Society* to the American publishers of *A History of Western Philosophy*, Russell wrote back in July 1946 requesting another copy, explaining that he needed to reread it but his books were inaccessible as he was moving house.

Popper did send another copy. And Russell, this time at least having given it his full attention, was most impressed. In a lecture entitled "Philosophy and Politics" and delivered in the same month as the meeting in H3, he proclaimed that an attack on Plato's political philosophy "has been brilliantly advocated in a recent book by Dr. K. R. Popper." In 1960 he advised a student to go to the LSE rather than Cambridge, because in London the philosophy was "vigorous."

The admiration was far more than just reciprocated. Popper believed Russell to be the most brilliant philosopher since Immanuel Kant and A *History of Western Philosophy* the finest overview of the subject ever written. In an address delivered on the Austrian Broadcasting Service in January 1947, he reviewed it in terms that to a non-Viennese would have seemed extravagantly effusive. Russell was described as the only great philosopher of the time, one who had been the most important contributor to logic since Aristotle. What made the book great, Popper rhapsodized, was the man. "Verging on hero-worship" is Peter Munz's characterization of Popper's attitude to Russell.

Russell's command of elegant and straightforward prose was a quality Popper particularly admired. When he began to write in English, he consciously tried to imitate Russell's style. By contrast, and in what may be a dig at Wittgenstein, he was scathing about the turgid "German" approach in which "every intellectual wants to show that he is in possession of all the ultimate secrets of the world." Russell was never deliberately obscurantist or pretentious, and in this respect Popper saw him as "our great master. Even when one cannot agree with him, one must always admire him. He always speaks clearly, simply and forcefully."

Popper remained awestruck at Russell's prodigious output, buying and reading almost all his books. There were some artists and writers, he wrote many years later, who were capable of creating a faultless work without any preliminary attempts: they achieve perfection immediately. "Amongst philosophers, Bertrand Russell was a genius of this kind. He wrote the most beautiful English; and in his manuscripts there was perhaps just one single word changed, in three pages, or perhaps in four pages."

In 1959 Popper requested and received permission from Russell to dedicate a book to him. In fact the work, the intended title of which was *Postscript after Twenty Years*, took years to be published. It eventually appeared as *Postscript to the Logic of Scientific Discovery*—in three parts—and by then Popper may well have forgotten his approach to Russell; the dedication did not appear. But he had initially proposed the following wording:

> *To Bertrand Russell*
> *Whose lucidity*
> *Sense of proportion*
> *And devotion to truth*
> *Have set an unattainable standard*
> *Of philosophical writing*

This was part of a very sporadic correspondence between Russell and Popper in the 1950s and 1960s. But Popper's hero worship was no impediment to Russell's getting the rough end of his pen over a refusal to review a volume of *Contemporary British Philosophy* to which Popper had contributed. Popper's letter has the tone of a resentful pupil trying to argue with his teacher. Russell's reply was conciliatory: "It had not occurred to me that you could take my refusal to do the review in question as in any degree derogatory to you."

Whatever Popper's hopes, he was never close to Russell. If he believed that his handling of Wittgenstein in H3 would bring him a little hero worship from Russell in return, the tactic was not a success. While Popper's writings are littered with references to Russell, in Russell's autobiography Popper does not merit a single mention.

6

The Faculty

Wittgenstein had an emancipatory effect.
—STEPHEN TOULMIN

THE LINEUP OF SPECTATORS in H3 is not yet complete. On the one hand there were the students, many of them Wittgenstein acolytes who walked and talked, dressed and debated like simulacra of their professor. On the other there were the dons. With one exception—John Wisdom—they were personally or professionally hostile to Wittgenstein. Their academic lives were founded on the existence of philosophical problems, the sine qua non of their careers. They taught philosophy much as it had traditionally been taught: Descartes and Kant, ethics and epistemology, philosophical logic and the philosophy of mind. That night their sympathies lay with the other Viennese philosopher in H3, Dr. Popper.

By 1946, Wittgenstein apart, the high noon of Cambridge philosophy had passed. Its twin native giants, Bertrand Russell and G. E. Moore, were entering their twilight years. In Russell's case, this twilight would be like the long half-light of a northern summer night; although already in his seventies, he still had a quarter of his life to live. Nevertheless, his best philosophy was far behind him. The same was true of Moore, his contemporary, who had embodied the intellectual and cultural elite of Cambridge before the First World War—the Cambridge that the young Wittgenstein had taken by storm.

Moore was now retired and guarded from intrusive visitors by his wife, Dorothy. He still turned up at Moral Science Club meetings on occasion, though not this time. Both Popper and Wittgenstein would have welcomed his presence—a fact that tells us more about Moore's character than about his philosophy. He was shy, attentive, and tolerant, and had a naive loyalty and unwavering integrity: altogether he possessed, in Russell's words, "a kind of exquisite purity." Russell once asked Moore whether he had ever lied. "Yes," replied Moore, which Russell believed to be his only untruth.

Popper had already had contact with most members of the Cambridge faculty, including Moore, who had both offered him a short-term lecturership in 1936 and then acted as a referee for his New Zealand post. Wittgenstein's relationship with Moore was long-standing and much closer. Three weeks after the meeting, when Wittgenstein responded to Popper's paper at the MSC, he sent Moore a letter. He would, he wrote, be honored if Moore were to come along. There is no evidence that Moore did. Mrs.

Moore tried to limit her husband's contact with Wittgenstein, who often left him exhausted.

Wittgenstein and Moore had first met in 1912, and the course of their relationship paints a vivid portrait both of Wittgenstein and of his relations with Cambridge. Moore was already famous, the young Austrian merely a student. Nevertheless, the older man was immediately taken by Wittgenstein, the only member of the audience to look puzzled during his lectures. He later wrote, "I soon came to feel that he was much cleverer at philosophy than I was, and not only cleverer, but also much more profound, and with a much better insight into the sort of inquiry which was really important and best worth pursuing, and into the best method of pursuing such inquiries."

The balance of power between them rapidly began to shift. In 1912 Moore vacated his rooms at the top of Whewell's Court, making way for Wittgenstein symbolically as well as literally. An indication of how far the balance had tilted was a journey Moore made to Norway in 1914, reluctantly and only at Wittgenstein's insistence, during which he was violently seasick. Wittgenstein had by now exiled himself to a small village north of Bergen and was leading a solitary existence, walking and thinking about logic. Once settled in, the don had as his principal task to take the student's dictation. Wittgenstein would then correct his notes, falling into a "terrible rage" with him when Moore failed to understand.

Upon his return, Moore asked the university authorities on Wittgenstein's behalf whether his treatise on logic would suffice for a Bachelor of Arts degree. He was informed that it would not: it had not been submitted in the appropriate format, with the requisite preface, footnotes, and so on. Moore conveyed the news to Norway. It prompted a letter of such force and brutality that relations between the two were severed: "If I am not worth your making an exception for me *even in some* STUPID *details* then I may as well go to HELL directly; and if I *am* worth it and you don't do it then—by God—*you* might go there." Moore was profoundly shaken and upset: he had been trying to help. The letter reverberated around his head for weeks. The two men did not talk again until they found themselves on the same train from London when Wittgenstein returned to Cambridge in 1929. The chance meeting led to their resuming a friendship of sorts.

Until Wittgenstein appeared, Russell had thought Moore fulfilled his ideal of genius. Yet Wittgenstein never had any regard for Moore's mental ability—Moore was, he thought, living proof of how far one could get in life with "absolutely no intelligence whatever." In fact Moore was a figure of internationally recognized stature and, with Russell, was revered as a pioneer of the analytic approach. Today's philosophy students are accustomed to lecturers greeting their rudimentary comments with the refrain "What exactly do you *mean?*" Moore should have patented this question; it was his catchphrase, and no day was quite complete without it being put. Moore insisted upon exactitude.

His breadth of interest was impressive. He made important contributions to debates over realism and idealism, certainty and skepticism, language and logic. A great advocate of common sense, he had once famously proclaimed that he could prove the existence of an external world by holding out his two hands and saying, "Here is one hand" and "Here is another." However, it was his book on morality, *Principia Ethica*, for which he was best known. When it was published, in 1903, it became an instant success and was adopted by the Bloomsbury Group as a sacred text—though one that was probably more skimmed than pored over.

Virginia Woolf asks in one of her letters, "Did you ever read the book that made us all so wise and good: *Principia Ethica?*"

In *Principia Ethica*, Moore argues that the "good" in ethics is essentially indefinable—rather like the color yellow. "Good is good," he wrote, "and that is the end of the matter." He gave the label "The Naturalistic Fallacy" to the mistake of trying to express goodness in other ways. It was an error similar to the one the eighteenth-century philosopher David Hume claimed was made whenever we try to derive an "ought" from an "is"—that is, to step from fact to value. One cannot logically move from a description of a state of affairs ("There are people starving in Burundi") to a moral judgment ("We should send them food"): the one does not *logically* follow from the other.

How, then, do we know what is the right thing to do? Moore believed that we access the good through intuition—intuition is the mind's moral eye. We perceive the good exactly as we see the color yellow. In place of our parents, our teachers, the state, or the Bible, our conscience becomes our moral authority. The Bloomsbury Group heard Moore's message as one of liberation, giving a green light to experimentation and sexual openness—or, as non-Bloomsberries might have said, to promiscuity.

IT IS DIFFICULT to imagine the 1946 Moral Science Faculty adopting such a message of liberation in their own lives, or in the pastoral advice given to their tutees. "Dull, dull, dull" is how Michael Wolff emphatically dismisses them. Those perfectly decent and conscientious, if unexceptional, academics did, however, serve a valuable purpose. It was said that "once a pupil went to Wittgenstein, he would henceforth have little time for other teachers." To this must be added a caveat: students "didn't necessarily come away from Wittgenstein with ability enhanced." For cerebral pyrotechnics, for the highest intellectual demands, Whewell's Court was the place to be. But it was to the "dull" men that students had to go if they wanted to pass exams.

Unlike Oxford, 1940s Cambridge had few academic philosophers. In Oxford, a relatively new course had been introduced— Politics, Philosophy and Economics. Its popularity grew rapidly, and each of the colleges had taken on its own philosophy don to meet the demand; the larger and richer of the colleges had two or three. Cambridge as a whole had to make do with half a dozen or so. Inevitably there was a shift in the philosophical center of power. In so far as there developed a Wittgenstein school in Britain, after the Second World War Oxford became its headquarters. In Cambridge, though some of the faculty acknowledged Wittgenstein's influence, even expressing gratitude to him in their books and articles, in practice their teaching of philosophy owed little to his approach.

There were four full-time philosophy dons: C. D. Broad, R. B. Braithwaite, J. Wisdom, and A. C. Ewing. All but Broad were present in H3. While Wittgenstein divided his life between Cambridge, Vienna, and Norway, his colleagues spent the bulk of their careers at the university. They had no disciples—and would probably have been deeply embarrassed to have attracted any. On Cambridge and on philosophy they left little mark—but that is the fate of most philosophy dons. In public, they exemplified the manners and deportment of English gentlemen—a world away from the loud Viennese expressiveness of both Wittgenstein and Popper. They valued highly the principle of tolerance; in debate, they believed in trying to see things from the other person's point of view. They spoke in courteous, measured tones, rarely raising a

voice in anger (though many of their students regarded such civ-
ilized attributes as stultifying). As they watched the rhetorical aggression of the H3 interchanges they must have felt awkward and
alarmed.

RICHARD BEVAN BRAITHWAITE, the tenant of H3, had a
minor part in the action, in Peter Geach's account, crawling
through the legs of the students to retrieve the poker. He was
among those who were acquainted with both protagonists. He
had first met Popper when the latter visited London in 1936, after
which Popper cited him as a referee for his grant application to
the Academic Assistance Council—a British body that offered financial support to academic refugees. It was to be the start of a
lifelong, though always arm's-length and formal, friendship. They
met again a few months after Popper's arrival from New Zealand

*Richard Braithwaite, King's
philosophy don and tenant of H3,
said to have spirited away the
poker Wittgenstein waved
at Popper.*

at the start of 1946, at a joint meeting of the Aristotelian Society and the Mind Association in the north of England, Braithwaite opening the proceedings and Popper delivering a technical paper on the status of the rules of logic. Braithwaite was Popper's main contact in Cambridge. Before the meeting, he had advised the guest on train times from London, invited him to dine at King's high table, and offered to put him up for the night at his home.

Born in 1900, Braithwaite was elected to a fellowship at King's in 1924 and was regarded as sufficiently superior to be invited to join the Apostles, the exclusive Cambridge secret society for the intellectually exalted. He had identified the importance of the *Tractatus* early on, reading a paper on Wittgenstein's book to the Moral Science Club in 1923. In 1953 he would become Knightsbridge Professor of Moral Philosophy—though ethics was never his speciality. In that role he would be one of the first to transfer to philosophy the tools developed by mathematicians and game theorists. Economists had already recognized the potential of using simple games to simulate complex human interaction; Braithwaite applied the same techniques to morality. In one hypothetical case, he imagined two bachelors, Matthew and Luke, who occupy adjacent flats. Luke likes to spend his evenings playing the piano; Matthew's hobby is the jazz trumpet. They each need peace and quiet in which to practice. Given various assumptions about their preferences, Braithwaite proved how the optimal solution is for Luke to play classical music for 17 evenings, to Matthew's 26 on the trumpet.

Such brain twisters were familiar territory to Popper. More importantly for their relationship, Popper and Braithwaite shared an interest in the philosophy of science, in probability, in infinity, and in causation.

Causation was of particular fascination in Cambridge—not just to Braithwaite, but to Broad and Russell too. They were all intrigued by a hypothetical case involving two factories, one in Manchester, the other in London. Each factory has a hooter, which signals the end of the morning shift at exactly twelve o'clock. It may then be empirically true that every time the hooter hoots at noon in Manchester the workers in London down tools. We would see, as Hume might have said, a contiguity of events—the northern hooter always being followed by the exit of the southern workers. Yet clearly the one was not the cause of the other. The question was, Why not? Wherein lay the difference between two events coincidentally linked and two events causally linked? How could one identify the mysterious power of causation—a furtive, cloak-and-dagger agent, never seen or touched? Perhaps causation was a chimera, a trick played on us by our imagination.

When Popper insisted that these were real philosophical problems, he could count on Braithwaite's support. But, even if Braithwaite had not been sympathetic to Popper's philosophical project, he had another reason to be on his side in H3. Thirteen years earlier he had been forced to make a public apology to Wittgenstein in *Mind*, the country's foremost philosophical periodical, read by all his peers.

This had arisen out of Wittgenstein's constant suspicion that he was being plagiarized. Braithwaite had attended the seminars Wittgenstein gave after his return to Cambridge from Vienna in 1929. In 1933 he then wrote an article in *University Studies* in which he tried to clarify some of Wittgenstein's evolving ideas. So enraged was Wittgenstein that he shot off a letter to *Mind*, disclaiming any link between his real views and those that Braithwaite "falsely" attributed to him. In response, Braithwaite wrote a contrite letter to *Mind* for taking Wittgenstein's name in vain. He ended, however, with a barbed remark: "The extent to which I have misrepresented Dr. Wittgenstein cannot be judged until the appearance of the book which we are all eagerly awaiting." Braithwaite might have suspected that Wittgenstein's relentless quest for perfection would prevent any such publication.

Braithwaite himself had no qualms about putting his thoughts in print. His 1946 lectures eventually appeared as a book, in which he wrote, "It is clear to me that I should not be philosophizing in the way I do had it not been my good fortune to have sat at the feet, in Cambridge, of G. E. Moore and of Ludwig Wittgenstein." Yet Wittgenstein barely merits a mention in the body of the text. And when Wittgenstein retired, in 1947, Braithwaite wrote to encourage an application for the vacancy from a possible candidate who wholeheartedly opposed Wittgenstein's method. In the event, Popper decided against applying: it had not improved his chances that he had once been rude to Professor Broad about the latter's fascination with the paranormal. Instead of Wittgenstein's chair passing to Popper, it was taken by G. H. von Wright, a devout Wittgensteinian.

There may have been another Braithwaite in H3 that night. It is said that among the audience was his idiosyncratic second wife, also known by her maiden name, Margaret Masterman. She was the daughter of the Liberal Cabinet minister Charles Masterman, who created Britain's World War One propaganda unit. As a former secretary of the MSC, she made a habit of showing up at meetings and seminars attended by her husband. Her usual practice was to sit on the windowsill. According to one, perhaps overimaginative, eyewitness, she was famed for not wearing knickers.

(He claims that he was distracted from the poker incident by a constant crossing and recrossing of her legs.)

The Braithwaites were hospitable and generous. They always offered to entertain the Poppers when they visited Cambridge. Beyond that, as we shall see, they stepped in when Wittgenstein turned his back on an old colleague of his, Friedrich Waismann, who had fled from Vienna. The Braithwaites gave the Waismann family shelter, a little money, and companionship.

ALMOST CERTAINLY PLAYING no part in the debate was another member of the faculty, Alfred Cyril Ewing, recollected by Michael Wolff as a "drab little man." If he spotted him among the crowd, Popper had cause to remember him with gratitude. It was Ewing who wrote to Popper in 1936 officially offering him a short-term lecturership in Cambridge, after the head of department, G. E. Moore, had agreed on the financial package with the Academic Assistance Council.

A year older than the century, Ewing had studied at Oxford, taught for several years in Wales, and arrived in Cambridge as a lecturer in 1931. The Reverend Maurice Wiles recalls Ewing's tutorials. "He was very methodical. He would talk for a bit and then say, 'I will now dictate.' You felt you were back in school. It was very depressing. He always had a worked-out answer to everything. There was no flexibility." Ewing wore heavy boots better suited to mountain climbing than to the flatness of East Anglia — as though "he was frightened of getting his feet wet." The mathematician Georg Kreisel described him as the sort of person "who looked like he still lived with his mother," as indeed he did.

He was deeply religious and serious. A. J. Ayer ribbed him about his belief in the afterlife, demanding to know what he most looked forward to in the next world. Ewing replied immediately, "God will tell me whether there are synthetic a priori propositions."

How much of the H3 debate Ewing followed is unclear. Maurice Wiles once admitted to Ewing that he did not understand a word of what Wittgenstein said. "Neither do I," confessed Ewing. Wittgenstein himself never bothered to hide his contempt for Ewing, even in front of students. A preoccupation of Wittgenstein's was solipsism––the theory that one can have secure knowledge only about oneself—and at an earlier MSC meeting he used Ewing as an illustration: "Let us make the purely hypothetical assumption that Ewing has a mind." More straightforwardly, Wittgenstein was damning about Ewing's work. In a discussion at Cornell, he quoted Ewing's definition "Good is what it is right to admire."

> Then he shook his head over it. "The definition throws no light. There are three concepts, all of them vague. Imagine three solid pieces of stone. You pick them up, fit them together and you now get a ball. What you've now got tells you something about the three shapes. Now consider you have three balls of soft mud or putty. Now you put the three together and mold out of them a ball. Ewing makes a soft ball out of three pieces of mud."

The final member of the Moral Sciences Faculty at the meeting, John Wisdom, was the one Cambridge philosopher to become a fully signed-up advocate of the Wittgensteinian method. Wisdom was popular, approachable, recognizably human, and, on the whole, diligent—though he would occasionally cancel classes so that he could cycle to the races at Newmarket to place bets and test his theories of probability.

Wisdom, like Braithwaite, had joined Popper a few months earlier at the Joint Session of the Mind Association and the Aristotelian Society. Then he had raised the question of how we know when a person is angry. Is it exactly like knowing that a kettle is boiling—which we deduce by its physical symptoms? Can anger too—a mental phenomenon, a feeling—be deduced only from its outward manifestations?

His detailed examination of the use of language, and what this reveals about the intricate and multifarious structures of grammar, used a method recycled from Wittgenstein. Before gaining a lecturership in Cambridge, Wisdom had spent several years at St. Andrews University in Scotland, but his arrival in Cambridge in 1934 marked a sharp break in his work and style. It was then that he began to attend Wittgenstein's seminars.

Like many of Wittgenstein's followers, however, he walked a tightrope between admiration and fear, attempting to please but at the same time not to presume. This is evident in the first essay in his book *Other Minds*, where he writes, "How much in this paper is due to Wittgenstein will be appreciated only by people who have listened to him. My debt to him is enormous. . . . At the same time I do not think that my way of doing things would quite meet with his approval—it's not sufficiently hard working—a bit cheap and flash." Nevertheless, he copied Wittgenstein's style and approach and shared his skepticism about what philosophy could actually achieve. First-year students who stuck their heads into the opening lecture of his course would be greeted with the question "Are you looking for wisdom in philosophy?"

ALTHOUGH HE WAS ABSENT on 25 October, and might any-way have been reluctant to attend the meeting had he been in Cambridge, the final member of the faculty, C. D. Broad, should have his place here. He was Braithwaite's predecessor as Knights-bridge Professor of Moral Philosophy, and the best known of the four. That autumn he was enjoying a sabbatical in Sweden, though several reports mistakenly locate him in H3.

Broad represented and molded the non-Wittgensteinian wing of philosophy in Cambridge, and had developed a considerable reputation beyond the university, based in part on the major works he produced in the 1920s and 1930s. These dealt with such perennial questions as the relationship between mind and body, how we can justify our knowledge of the external world, and what occurs in the mind when one has a perception of an object. By 1946 Broad's attention had shifted to ethics. In an essay written shortly before the meeting, he had considered the ethics of a ter-rorist action that might affect innocent bystanders as well as the intended victim. Wittgenstein never expended energy in the analysis of such practical ethical issues. For him, morality always remained one of those areas which could be shown but not com-mentated upon, being revealed in the way people conducted their lives but not susceptible to logical rigor.

"Reliable rather than brilliant" had been Russell's early and perceptive assessment of Broad when, in Wittgenstein's shadow, he had been Russell's pupil. As a teacher, he had donnish foibles that are the stuff of reminiscence at reunion dinners. He used to script his lectures fully in advance, and then read each sentence aloud, twice. The jokes he read out three times. That, says Maurice Wiles, who attended Broad's lectures, was the only way one could tell what was a joke. When his course was interrupted by a term's sabbatical, Broad began the first lecture after his absence with "Point D . . ."

Though Broad was painstaking and dreary in the lecture theater, he enjoyed malicious gossip outside, incessantly carping about Wittgenstein behind his back and sprinkling his writings with snide references. He admitted that he disliked attending the Moral Science Club. He was not "quick-witted nor quick tongued enough to take a useful part in philosophical discussion by word of mouth; and I was not prepared to spend hours every week in a thick atmosphere of cigarette-smoke, while Wittgenstein punctually went through the hoops, and the faithful, as punctually, wondered with a foolish face of praise." In a book published in the mid-1920s, Broad remarked on "the philosophical gambols of my younger friends as they dance to the highly syncopated pipings of Herr Wittgenstein's flute."

The uneasy nature of the relationship between Broad and Wittgenstein persisted until the end. Joan Bevan, the wife of the doctor who took Wittgenstein in before his death, once played a practical joke on her guest, telling him that Broad was coming round for tea. When he discovered the truth, Wittgenstein went into a deep sulk, refusing to speak to his hostess for two days.

Nevertheless, it was a mark of Broad's overriding sense of justice—a trait Wittgenstein (and Popper) appreciated and always associated with the English—that he backed Wittgenstein's appointment to the professorship when Moore retired in 1939. Broad was quoted as saying, "To refuse the chair to Wittgenstein would be like refusing Einstein a chair of physics." He also came down on Wittgenstein's side in a bizarre dispute during the war about whether Wittgenstein should be paid. Wittgenstein insisted not.

At the time, 1942, Wittgenstein was working as a dispensary assistant at Guy's Hospital, London, returning to Cambridge at weekends to lecture. These classes, he felt, were unsuccessful — perhaps, in the middle of the war, because of the caliber of students. He therefore proposed changing his lectures to "at-homes," for which he wanted to be taken off the payroll and given only expenses until he was content that the new format worked. Broad, who had taken on an additional job as a bursar at Trinity — describing this as his contribution to the war effort — wrote to the Moral Science Faculty noting that Wittgenstein was

> an intensely conscientious man with a very exalted standard; and I have no doubt that most of us would have no hesitation in accepting payment for what he will do. Still the fact remains that he feels intensely uncomfortable. He cannot help continuing to philosophize at every possible opportunity and for him it is an essential part of philosophizing to carry on a kind of Socratic dialogue with the class.

And Broad was confident that Wittgenstein would be completely honest. "Knowing Wittgenstein I am quite sure that the University runs no risks under such an arrangement."

7

A Viennese Whirl

*I realize that relations between the Viennese philosophers
of the early 1920s were complex, full of stress, and often
paranoid.*

—STEPHEN TOULMIN

TO AN OUTSIDER, a violent confrontation between Wittgenstein and Popper might have seemed implausible. Superficially, they had in common a civilization—and its dissolution. Although Wittgenstein was the older by thirteen years, they had shared the cultural excitement and cosmopolitan politics of the last years of the Austro-Hungarian Empire. They had in common, too, the impact on their lives of the lost First World War, the attempt to raise a modern republic on the ruins of the monarchy, the descent into the corporate state, and the maelstrom of Hitler and Nazism.

And, of course, Vienna. There, in the marbled halls of the Palais Wittgenstein at Alleegasse 16, lived the Austrian steel magnate Karl Wittgenstein. Ludwig Josef Johann, born at 8:30 in the evening of 26 April 1889, was his eighth and last child. Just a mile away, looking down to the south door of St. Stephen's Cathedral, was the comfortable, book-lined apartment where Karl Raimund Popper, born on 28 July 1902, would grow up. He was the youngest of three children of a well-to-do lawyer. Between the two homes rose the seat of Habsburg rule, the Hofburg, where Emperor Franz Josef, "the first bureaucrat of the empire," would more likely than not have been busy at work in his plainly furnished office.

For an imperial capital that at its zenith ruled over Hungarians, Czechs, Slovaks, Poles, Italians, Galicians, Slovenes, Serbs and Croats—and Austrians—Vienna was a surprisingly tight-knit city. With their Jewish origins, their interest in music, their contacts with cultural radicals, their training as teachers, and their connections with the fountainhead of logical positivism, the Vienna Circle, Wittgenstein and Popper had many potential links. That the intersecting cultural, social, and academic circles in which they moved never resulted in their meeting is remarkable. Wittgenstein knew the architect Adolf Loos, who knew Arnold Schoenberg, to whose private music-society concerts Popper went. And it was common knowledge where, within the tight confines of the Ringstrasse, all Vienna's epic figures were to be found, and when. This was the world of the coffeehouse and the *Stammtisch*, the regulars' table. Over a coffee, a glass of water, and perhaps a strudel, an article would be written, an argument renewed, a play reviewed, an introduction made.

Want a word about a modernist building project with Loos or about twelve-tone music with Alban Berg? Try the Café Museum or perhaps the Herrenhof. Looking to pick a bone with Karl Kraus over one of his coruscating articles in *Die Fackel?* He gives permission to meet him in the Café Central in the evening, when he eats his dinner—a very sharp sausage. Remember, he works all night and sleeps into the day. You can meet Peter Altenberg, the poet, there too. Probably he'll be writing one of the stream of postcards with which he keeps in touch with his friends. The mathematicians, like Gödel, are to be found in the coffeehouses with the white tabletops, on which they scrawl equations. Fancy a game of chess? Try your luck against the political refugee and coffeehouse habitué Lev Bronstein—later to become better known under his revolutionary alias, "Trotsky." If you are after a crime reporter from a popular paper you might have to go down-market a bit—somewhere like Joseph Roth's Café Wirzl, with its "grimy playing cards and the smell of coffee, Okocimer beer, cheap cigars and bread sticks," where the reporters consult tarot cards while waiting for their sources to arrive.

Putting aside the diverting image of Wittgenstein and Popper playing cards and drinking beer together at the Wirzl, it is clear that they had many friends and acquaintances in common in Vienna, and must often have been in close proximity. That was certainly the case on 15 July 1927, when the police opened fire on demonstrating Social Democrat workers and bystanders, killing eighty-five. "My wife-to-be and I were among the incredulous witnesses of the scene," wrote Popper. Somewhere in the vicinity were Wittgenstein and his sister Margarete. At his insistence, she had sent her car and chauffeur away and was walking with him. When she wanted to turn tail at the shooting, he told her sternly, "When one hears rifle fire, one doesn't run."

As for their intersecting social lives, one example was their families' relationships with the Freuds. Sigmund Freud's sister, Rosa Graf, was a close friend of Popper's parents. In 1916 she was on holiday with the Poppers when her son, in uniform, made what was to be his last visit to her; shortly after, he became a casualty of the war. Wittgenstein's sister Margarete, who was on the fringes of many of the disparate intellectual and artistic movements flowering in Vienna, became acquainted with Freud in the early 1930s. After the First World War she was appointed by the American Relief Administrator, and future president of the United States, Herbert Hoover, to be Special Representative of the American Relief Program for Austria. Then she worked in juvenile prisons and in Graz University as a psychotherapeutic adviser, which brought her to Freud's shawl-draped couch. To enlarge her understanding of the treatment of neurosis, she was analyzed by him for two years, and they remained in close contact until his death. On 3 June 1938—the day he fled Vienna—he inscribed a copy of his book *The Future of an Illusion*: "Mrs. Margaret Stonborough on the occasion of my temporary departure."

Freud's work touched both her brother and Karl Popper intellectually, but to quite opposite effect. Wittgenstein drew parallels between his own later work and psychotherapy. Popper targeted Freudianism as a particularly flaccid specimen of pseudoscience.

THE REFORM OF EDUCATION was another aspect of the city's cultural ferment which had a direct bearing on their philosophical development. Both Popper and Wittgenstein trained as teachers in Vienna—and within four years of each other. Both spent time teaching children—Wittgenstein in primary schools in the Austrian countryside, Popper in primary and secondary schools and, under the aegis of the psychiatrist (and former colleague of Freud) Alfred Adler, with disadvantaged children in Vienna. Both came under the influence of the exuberant Karl Bühler, Professor of Philosophy at the Pedagogic Institute, and of Otto Glöckel, Vienna's Councillor for Education and the moving spirit of a short-lived experiment in Austrian schooling. Glöckel is thought to have been in close touch with Margarete Stonborough when she was Hoover's representative.

Wittgenstein turned to teaching after his release from an Italian prisoner-of-war camp in 1919. This change of direction was no passing fad; he taught for some six years in villages deep in the countryside—an episode that must be understood in the context of his family's long-standing commitment to social work. His eldest sister, Hermine, also became involved with the education of the poor. And Margarete stated in her 1942 application for a job with the American Red Cross (which mysteriously came to rest with the Office of Strategic Services, the forerunner of the CIA, when it was considering offering her a post) that she had worked throughout her adult life and had never been paid. But, unlike his siblings, Ludwig was not simply responding to an aristocratic sense of charitable obligation: he now wanted to strip his life of all unnecessary frills and fripperies, to deprive himself of any hint of comfort, let alone luxury. He became an ascetic, bringing his schoolteaching to the children of Austria's rural poor, working in remote areas, in villages finally reachable only on foot.

For Karl Popper the move into education was not a question of asceticism: it followed naturally from his work with disadvantaged children, which he began after leaving school. Another strong motive for enrolling in Vienna's new Pedagogic Institute was that it shared some courses with the university, thus enabling him to transfer into the higher education which had been out of his reach because he had been unable to take his final leaving examination—the *Matura*. As we will see, his father's sudden impoverishment forced him to leave school early.

The Pedagogic Institute had been established to further the Austrian educational reform program. This attempted to steer education away from a "drill school" approach, in which schoolchildren were treated as empty vessels to be filled by the accumulation of dictated knowledge and respect for authority, toward seeking children's active engagement through self-discovery and problem solving. Both Popper and Wittgenstein were trained in the methods of encouraging this. Integral to the vision was a general view of the mind as innately capable of producing frameworks within which information could be organized.

Although Wittgenstein poked fun at the program's "more vulgar slogans and projects," his dictionary for schoolchildren, *Wörterbuch für Volksschulen*, using the dialects of the Austrian countryside and respecting its culture, was well within the spirit of the reforms. It was equally part of the philosophical project that led him toward the notion, in the *Philosophical Investigations*, that communities might use language in manifold perfectly valid ways. It could also be seen as informing his mode of teaching, conjuring up examples and interrogating the students' responses.

Popper's training brought him a formative encounter with Karl Bühler; he took from Bühler the view that we think in terms of problems and their tentative solutions. This, Popper would later argue, is how science progresses. Scientists do not assemble facts and see what these add up to: rather, they imagine solutions and only then search for the evidence to support them.

Bühler had been invited to Vienna by Glöckel. By then Wittgenstein was already deep in the countryside. However, though he was never a student of Bühler's, the language-learning processes of a child were a matter of absorbing interest for him and Wittgenstein evidently knew both the man and his work. (Bühler and his wife, Charlotte, an eminent child psychologist, were present at the crucial first meeting between Moritz Schlick and Wittgenstein, arranged by Margarete.) From time to time Wittgenstein would denounce Bühler as a charlatan.

Popper took the opposite view: "From the teachers in the Pedagogic Institute I learned very little, but I learned much from Karl Bühler." The appreciation was mutual. In a reference Bühler gave Popper for his lectureship in New Zealand, he described Popper's doctoral thesis as "a very sagacious philosophical investigation." He added, "I am highly estimating [sic] his abilities as a teacher."

Yet these links between the two Viennese also point toward a telling divide. On one side we see the chauffeur, the voluntary charitable work, the unforced decision to leave Vienna for the impoverished countryside. On the other stands bare necessity. To understand the depth of the divide, a visit to Wittgenstein's home in the Alleegasse is required.

8

The Concerts in the Palais

The multimillionaire as a village schoolmaster surely is a piece of perversity.

—THOMAS BERNHARD

BOTH POPPER AND WITTGENSTEIN came from highly cultivated backgrounds. Popper's father was a lawyer whose apartment and office were in the heart of Vienna. He had a library of ten thousand books, and as a hobby he translated the Greek and Roman classics into German. He was also concerned for the homeless, sitting on committees to provide impoverished working men with housing—one of their hostels sheltered Hitler during his early years in Vienna. For his work, he received an imperial decoration, being made a Knight of the Order of Franz Josef. But the Wittgensteins were in another class—one from which they unreservedly looked down on bourgeois families like the Poppers.

By the end of the nineteenth century the Wittgensteins had taken their place among the Austrian superrich, second only to the Vienna branch of the Rothschild family. The prime force in his country's steel cartel, able to bend the price of steel at will, Karl Wittgenstein was a business genius. It was said that, if he had been German, Bismarck would have brought him into the management of the economy. It would have been like offering Carnegie, Mellon, or Rockefeller a place in the American administration.

His home was the magnificent Palais Wittgenstein, in the Alleegasse, now the Argentinierstrasse (the site is occupied by a rundown postwar block of flats). Opposed to parading his family's riches, Karl Wittgenstein shunned the title "Palais": to him, it was "Haus" Wittgenstein. It stood close by the massive baroque grandeur of Charles VI's imperial church, the Karlskirche, in the heart of the area colonized in the late nineteenth century by the new aristocracy of commerce and industry. Here were the opulent residences of families who now stood just one step below the established nobility of court and government in the stiff and fusty hierarchy of Austro-Hungarian society. Brahms said of the Wittgensteins, whom he visited regularly, "They all seemed to act with one another as if they were at court."

There were public obligations attached to such status. The Wittgenstein house was one of the preeminent musical salons in the city of Mahler, Schoenberg, Webern, Berg, and of course, Brahms. The first performance of Brahms's Clarinet Quintet took place there. The composer gave Karl and Leopoldine's musically talented children piano lessons, and once rubbed vintage champagne into Margarete's scalp to make her hair grow again after it had been cut short when she was ill. Clara Schumann, Mahler, and the conductor Bruno Walter were frequent guests too. (Walter was a relative of Popper's grandmother.) Richard Strauss played duets with Ludwig's brother Paul, a concert pianist who lost his right arm in the First World War and for whom, in 1931, Ravel wrote his Piano Concerto in D for the Left Hand. (Paul rejected a work for the left hand he had commissioned from Prokofiev: "I do not understand a note of it and I shall not play it." Prokofiev retorted that, musically, Paul belonged in the last century.) It would not be too much of an exaggeration to say that, while the Poppers went to the concerts they supported, the concerts and recitals the Wittgensteins patronized came to them, where pianists had a choice of six grand pianos on which to perform.

Bruno Walter, who was assistant conductor of the Vienna State Opera from 1901 to 1912 and later its musical director wrote in his memoirs, "The Wittgensteins continued the noble tradition of those leading Viennese groups who considered it incumbent upon them to further art and artists. The Wittgenstein house was frequented by prominent painters and sculptors and by leading men from the world of science. I always enjoyed with gratification the all-pervading atmosphere of humanity and culture." There was, nevertheless, an ambiguity in the Wittgensteins' relationship with the old nobility—a keeping of the family apart that ran together with the wish to be unobtrusive. This manifested itself in Karl Wittgenstein's insistence on "Haus Wittgenstein" and in the anonymity of his huge charitable donations. Karl refused Ludwig's sisters riding lessons, so that they should not grow up think-

ing of themselves as aristocratic. And when a nobleman was made
Minister of Finance, Karl published an attack on the appointment, arguing vigorously that being a count was not a sufficient
qualification.

Karl saw himself as a radical, and as such he was a major force
in support of a revolution in the visual arts: in 1897 it was largely
his money that had financed the Secession Building for artists
who had broken away from the deadening approved school of
grand subjects, grandly treated. The painter Gustav Klimt called
him "the minister of fine art" and painted a portrait of Margarete
on her marriage in 1905. In the luxuriant eroticism of the compo-
sition, her dark eyes give a hint of unease. As soon as possible she
hid the picture away in the attic of her country house.

Although the Wittgensteins may have tried to live inconspicu-
ously, their wealth and patronage of the arts were not to everyone's
taste. The periodical *Die Fackel* lampooned Vienna's leading
families who prided themselves on their generous benefactions.
And Thomas Bernhard, Austria's finest contemporary novelist and
playwright, whose work demonstrates an obsession with Ludwig,
continues this strain of invective against the rich. In his fictional-
ized memoir *Wittgenstein's Nephew*, first published in 1982, he
makes a savage comment on the Wittgensteins' patronage of
Klimt. Bernhard targets their

> repulsive paintings from the Klimt period, including one by Klimt
> himself, by whom the arms-manufacturing Wittgensteins had
> themselves portrayed, as indeed also by other famous painters of
> the day, since it was the fashion among the so-called newly rich of
> the turn of the century to have themselves painted under the pre-

*Wittgenstein and his sisters (left to right) Hermine, Helene, and
Margarete. Hermine was a mother to him. Luki is on the right;
his elder brother, Paul, is on the left.*

tence of Maecenas-like patronage. Basically the Wittgensteins,
like all the rest of their kind, had no time for the arts but they
wanted to be Maecenases.

Bernhard goes on to describe the family as "hostile to the arts and
the intellect, stifled by their fortune, by their millions."

AT LEAST BEFORE the First World War, Ludwig gave every
sign of enjoying his father's wealth. His Cambridge friend David
Pinsent, who himself came from a comfortably well-off background, expressed surprise to his diary when Wittgenstein suggested setting off on a holiday to Iceland to be paid for by his
father. "I asked what he estimated the cost would be: upon which

Popper and his sisters (left to right) *Dora and Annie. Karl was the baby of the family.*

he said—'Oh, that doesn't matter: I have no money and you have no money—at least, if you have, it doesn't matter. But my father has a lot.'—Upon which he proposed that his father should pay for us both." Then, when their journey began, there was the question of where Pinsent would stay in London. Wittgenstein took him to the Grand Hotel, Trafalgar Square: "I tried to suggest some less pretentious hotel—especially as Wittgenstein is staying with Russell in any case—but he would not hear of it. There is to be no sparing of expense on this trip." And while Wittgenstein eventually became famous for the spartan furnishing of his Cambridge rooms, that was not the case before the First World War. Pinsent records how in October 1912 he helped Wittgenstein move his furniture into his Trinity rooms. The furniture came from London: Wittgenstein had rejected what Cambridge had to offer as "beastly." "He has had his furniture all specially made for him on his own lines—rather quaint but not bad." On their return from Iceland, "We dined—in style with Champagne."

Wittgenstein's father died of cancer in 1913. Ludwig was said to have then become the richest man in Austria and one of the richest in Europe. While Popper's father lost all his savings in Austria's postwar inflation, Wittgenstein's father had protected much of his family's fortune by holding it abroad.

But Ludwig's wealth was fleeting. The war had changed him spiritually. His sister Hermine records soldiers referring to him as "the one with the Gospel," because he always carried Tolstoy's edition of the Gospels. On his return from captivity he transferred all his money to his remaining brother, Paul, and his sisters Hermine and Helene. (Margarete had married an affluent American, Jerome Stonborough, and was well provided for.) Hermine records the agonies Ludwig went through with the despairing notary to assure himself that he had put his fortune irretrievably out of reach. However, she also records that an essential part of his outlook was "his completely free and relaxed acceptance of the fact that he was ready to let his brother and sisters help him in any future situation."

It was from this time on that Wittgenstein's life took on an obsessively austere quality—combined with a passion for tidiness and cleanliness. His friend and later architectural collaborator Paul Engelmann puts this down to an

> overpowering urge to cast off all encumbrances that imposed an insupportable burden on his attitude to the outside world: his fortune as well as his necktie. The latter (I remember having been

told) he had in his early youth selected with particular care, and no doubt with his unerring taste. But he did not discard it to do penance. . . . [He] decided to shed all the things, big or small, that he felt to be petty or ludicrous.

Wittgenstein gave a similar explanation to his nephew John Stonborough. " 'If you were going for a long hike up a steep mountain,' he said, 'you would deposit your weighty rucksack at the bottom.' That was my uncle's attitude to money. He wanted to release himself from a heavy burden." In his obituary, *The Times* recorded that "Wittgenstein showed the characteristics of a religious contemplative of the hermit type," and referred to his extreme abnegation and retirement.

The conveniences of coming from a rich family were not discarded altogether. In the 1920s and 1930s, when he held conversations with the moving force of the Vienna Circle, Moritz Schlick, and one of its members, Friedrich Waismann, he had at his disposal various Wittgenstein houses where peace and quiet might be found. There was Neuwaldegg, in a suburb of Vienna and used by his family as a retreat in spring and autumn. There was a house in the Augustinierstrasse belonging to his brother and sister, where he held meetings in the unused office. And always there was the family summer retreat at the Hochreith, an hour's drive west of Vienna, deep among the hills. The attachment to Vienna and his sisters remained strong. From 1929, when he returned to Cambridge, until 1937, and from 1949 until his death in 1951, Wittgenstein regularly spent his summer and Christmas holidays in Austria.

A patrician manner was not so easily sloughed off as wealth.

The entrance to Alleegasse 16. To most Viennese its grandeur unmistakably merited the title "Palace." Karl Wittgenstein preferred the more unassuming "House."

Leavis saw Wittgenstein as a troubled soul. But he also saw that as being bound up with the philosopher's highborn bearing. "I suppose I am not the only one who thought of the quality I have called assurance as having, going as it did with cultivation and quiet distinction, something aristocratic about it." The angst Leavis perceived presumably sprang from the clash between plutocratic poise and the will to an austere lifestyle. Bernhard put it more harshly: "the multimillionaire as a village schoolmaster surely is a piece of perversity."

CULTIVATION CERTAINLY, perhaps a quiet distinction, but Popper had no aristocratic mien to set him apart, nor was there any family money to draw upon. In 1919–20 he too was living an austere life, though not by choice. He had left home to live in "a disused part of a former military hospital converted by students into an extremely primitive students' home. I wanted to be independent and I tried not to be a burden to my father, who was well over sixty and had lost all his savings in the runaway inflation after the war."

Although Karl Wittgenstein had insisted that his family should not flaunt their riches, in so tight-knit a city as Vienna the Wittgensteins would still have been well known to other Viennese families like the Poppers. The Wittgenstein name was news—and not only in the social and business pages of the daily press but in Karl Kraus's journal *Die Fackel*, in which he savaged the establishment with trenchant comment and satire. It is inconceivable that Karl Wittgenstein's business and charitable affairs, his articles on economics, and the family's position in Viennese cultural life did not come up at the Poppers' dinner table.

That Karl Popper did feel some personal animus is evident from a contemptuous remark, recalled by Peter Munz, that Ludwig Wittgenstein couldn't tell the difference between a coffeehouse and a trench. Coffeehouses had definite associations for Popper: they represented a life of ease for the affluent, frivolous time-wasting, fashionable thinking. He commented to one of his former students and later colleagues, the Israeli philosopher Joseph Agassi, that "the *Tractatus* smelled of the coffeehouse."

Popper was simply wrong on Wittgenstein's knowledge of the trenches. If the *Tractatus* smelled of anything, it was of death and decay. Wittgenstein had fought for Austria with conspicuous bravery as a volunteer in the First World War. He had used his family's social connections not to avoid combat but instead to obtain a posting to the front, when an operation at seventeen for a double hernia would have allowed him to remain far from the sound of gunfire. He took on the job of artillery forward observation officer, and insisted on holding his position long beyond the requirements of duty. It is said that he would have won the Austro-Hungarian equivalent of the Victoria Cross, but the battle in question was lost and no medals were awarded for defeats. And still, throughout the war, he continued to work on the *Tractatus*.

Paul Engelmann records that "Wittgenstein considered his duty to serve in the war as an overriding obligation. When he heard that his friend Bertrand Russell was in prison as an opponent of the war, he did not withhold his respect for Russell's personal courage, but felt that this was heroism in the wrong place."

In the Second World War, Wittgenstein again displayed a sense of obligation. Though over fifty, he arranged to leave Cambridge to work as a dispensary aide in a south-London hospital during the Blitz. Here too he displayed his gift for giving himself utterly to whatever task he had undertaken, assisting a medical team investigating wound trauma. When the team moved to Newcastle, he accepted their invitation to go with them. He might also have made another, though perverse, contribution to Britain's war. In 1939 he discussed contradictions in mathematical logic with Alan Turing, who thought Wittgenstein's view, that contradictions were not significant, utterly wrongheaded. (Wittgenstein's philosophy of language had evolved dramatically since the *Tractatus*. Then he had believed in a perfect, ideal language, devoid of ambiguity. Now he believed that if communities developed or adopted a language that contained internal contradictions, well, so be it.) The memory of that disagreement could have played a part in Turing's thinking on the logical design of the Bombe, the primitive computer, which made possible the timely cracking of the German Enigma code in Bletchley Park.

Despite his derogatory comments about Wittgenstein, Popper himself never fought in battle. He was only sixteen years old when the First World War came to an end; the Second he spent working several thousand miles from the front line, in the safety of New Zealand, from where he helped to organize the escape of some forty Austrian refugees. He tried to join the New Zealand armed forces but was turned down for physical reasons. However, he saw his contribution to the defeat of Nazism as being the writing of *The Poverty of Historicism* and *The Open Society and Its Enemies*. And so it was—even though publication of *The Open*

The Popper apartment on the second floor, looking down at St. Stephen's Cathedral, and home to ten thousand books.

Society was delayed beyond the fall of Hitler's Germany. He remarked in *Unended Quest* that these "were my war effort." In 1946 he told Isaiah Berlin and A. J. Ayer, in the presence of Ernest Gellner, that *The Open Society* "was a fighting book." It had been a fight in which Popper and Wittgenstein, who both came from Jewish families, had a personal interest.

9

Once a Jew

In Western civilization the Jew is always measured on scales that do not fit him.

—WITTGENSTEIN

WHATEVER THE DIFFERENCES of social standing and fortune between Wittgenstein and Popper, they shared one ineradicable characteristic: they belonged to assimilated Jewish families in the most assimilated city in Europe. And one way of seeing the Cambridge confrontation is as a clash between two exiles of Jewish extraction still rooted in Vienna. Yet a distinctive social and political culture, far from uniting them, showed only how dissimilar they were in their approach to life.

The issue of Jewish identity presented complex problems. The oppositional concepts of exclusion and assimilation cannot do justice to the position of the many Jews who occupied a permanently transitional place in Vienna's multinational Christian society under Franz Josef—neither fully excluded nor fully assimilated. On the Jewish side, legal emancipation led to an array of possible self-definitions. But, whatever the self-definition, the degree of social acceptance was always for someone else to decide. Exclusion, discrimination, unspoken reservations, the "Jewish Question"—all these were in the hands of the non-Jewish Christian majority.

Sigmund Freud could and did acknowledge a strong Jewish identity. "You no doubt know that I gladly and proudly acknowledge my Jewishness, though my Jewishness, though my attitude towards any religion, including ours, is critically negative," he wrote. But such a statement would not have been possible for Wittgenstein and Popper. Both came from among the many Jewish families that had been baptized into the Christian faith—in Popper's case, not long before his birth. Popper's elder sisters were born into a Jewish household.

Vienna's Jews, however, whether observant or simply Jewish by descent, tended to form a coherent and cohesive community, living, working, socializing, marrying within the same broad group. In turn-of-the-century Vienna, converted Jews still felt at home in the predominantly Jewish districts of Innenstadt, Leopoldstadt, and Alsergrund and found most of their friends among other Jewish families, converted or not.

Vienna had the highest conversion rate of Jews to Christianity of any European urban center—the result of both an internalization of the pervasive culture of anti-Semitism and a conviction that conversion was a necessary means of advancement in Habsburg society. Marriage laws which prohibited a union between

Jews and religious Christians also played a part. A marriage between those belonging to different religions required conversion
of one of the partners to the other's faith—or at least a declaration
of no faith by one of them. In a marriage between a Jew and a
non-Jew, the Jewish partner usually made the move.

Jews brought up in German-speaking culture and society
placed a high premium on assimilation, but full assimilation was
never feasible. Their German acculturation might be complete in
all aspects of family, working, cultural, and political life. But, as
the Viennese playwright and novelist Arthur Schnitzler remarked,
"It was not possible, especially not for a Jew in public life, to ignore the fact that he was a Jew. Nobody else was doing so, not the
Gentiles, and even less the Jews." This is not a purely Viennese
phenomenon. Alan Isler puts it pithily in his New York novel *The
Prince of West End Avenue:* "For the goyim he remains a Jew, of
course; and for the Jews, in view of his success, he is still a Jew
anyway." Often asked for his reactions "as a Jew," Popper would
angrily have recognized the truth of that.

There were many ways—subtle and not so subtle—in which
Jewish descent, conversion, and the manner of conversion could
be alluded to. The German historian Barbara Suchy has recorded
several such expressions. *"Liegend getauft"* (baptized as a baby)
was a nod-and-wink reference to someone's Jewish origins. The
composer Felix Mendelssohn was *"als Kind getauft"* (baptized as
a child). He was a friend of Ludwig's paternal grandmother,
Fanny Figdor, and the first sponsor of her nephew, the virtuoso vi-
olinist Joseph Joachim. More "Jewish," and so more alien than either of these possibilities in the eyes of those who used such
terms, was *"Übergetreten,"* meaning having deliberately decided
to convert.

In later years, some Jews would appropriate the terms for themselves. *"Liegend getauft"* was used, with a slightly mocking undertone and sometimes even a dash of *Schadenfreude*. You could say: *"das hat ihm auch nicht viel genützt"*—it didn't do him much good either. Or, in contrast to those who converted, it could be said: one should not blame him for deserting the community because it was not his decision to grow up as Catholic or Protestant. If the person involved is a big and famous name, a cultural hero, he or she is of course claimed as "one of us," entered into the proud listing of *Grosse Juden der Geschichte* (the Great Jews of History).

Popper might well have been referred to as *liegend getauft*, an innuendo that Wittgenstein was probably spared, since baptism in his family went back much further.

In Vienna, some sense of exclusion or alienation from the majority Christian society was the lot of many of the converted as well as the practicing Jews. Certainly Popper struck others as feeling the odd man out in the 1920s when he sought relaxation by attending Arnold Schoenberg's subscription concerts. A pupil of Schoenberg's, Lona Truding, remembered Popper then as "a wonderful man, as great a man as he is a thinker. He didn't fit in. He was an outsider in the best sense of the word." No doubt he always manifested a critical distance. The historian Malachi Hacohen puts it at a fundamental level: "The life and work of this Central European exile embody the dilemmas of liberalism, Jewish assimilation, and Central European cosmopolitanism."

In Wittgenstein, too, the feeling of being divorced from the world around him was part of his nature—but with a difference. He was born into the easy social acceptance that accompanies supreme wealth. He may have become consciously self-denying after the First World War, but Theodore Redpath saw him as always aware of his inherited position "as the scion of a wealthy upper-class Austrian family, and he sometimes surprised one by making this quite plain, as for instance his frequent use of the term 'Ringstrasse' for things he considered to be second-rate." The Ringstrasse was, and is, the grand, bustling avenue encircling central Vienna, but to Wittgenstein it denoted a place of pomp and gesture, empty of content. Even though the area enclosed by the Ringstrasse was the fashionable quarter, for a Wittgenstein the name was obviously not synonymous with real quality. In the same condescending spirit, toward the end of his life he condemned the ball gowns of young ladies going to a Trinity College May Ball as "tawdry"—not up to the standard of a reception at the Palais Wittgenstein, perhaps, in those glittering days before the First World War.

All this grandeur was comparatively recent. The social mobility of the Wittgensteins—a German Jewish family from Hesse—presents a case study in the tolerance of Franz Josef's realm. Born into the family of a minor German prince's estate manager, Ludwig's grandfather became a wool merchant, then a Viennese property dealer; his son, Ludwig's father, made himself an industrial tycoon, a benefactor of the arts, and an associate of ancient aristocracy—and all within eighty years. Yet, as the end of the 1930s was to reveal, the whole social edifice was built on the thinnest of Austrian ice.

THE VIENNA OF POPPER'S and Wittgenstein's formative years was the seedbed for Hitler and the Holocaust—what Karl

Kraus saw in a nightmare vision as "a proving ground for world destruction." The novelist Herman Kesten saw it as "a kind of vanished fairy tale Wild East." It was a city of "brilliant creation in a nonetheless decaying culture." That brilliant creation was the intellectual and cultural future: the new struggling to escape the suffocation of the old.

The origins of this revolution lay in the upheaval wrought by rapid industrialization in the nineteenth century—a revolution in which Karl Wittgenstein was a major force. By the turn of the century a novel outlook was emerging that rejected the certainties of the Enlightenment and the love of decoration and the obedience to tradition that weighed down imperial society, restricted its horizons, and stifled innovation. In their place came a demand for experiment, function dictating form, honesty, and clarity in expression.

Under the walls of the Hofburg but far removed in spirit from its dominating formality and heritage, this was the city of Ernst Mach and the theory of the fluctuating and uncertain self; of Freud and the power of the unconscious; of Schoenberg and the ousting of conventional tonality in favor of the twelve-tone system. Here within a single period were Arthur Schnitzler's literature of the interior monologue and of the sexual drive as the prime mover of human relationships; Adolf Loos and the stripping away of ornament for ornament's sake in architecture; Otto Weininger, the self-hating Jew, whose book *Sex and Character* Wittgenstein read as a young man and admired; and Karl Kraus and his attack on the linguistic forms—clichés, metaphors—that disguised realities in politics and culture. Kraus's demand that the language of public life should be cleansed of cultural dishonesty paralleled Wittgenstein's linguistic preoccupations.

It was also a city where intellectuals of Jewish descent played a dominating role, dynamically assimilating to its cosmopolitan complexion. Six of the leading figures named in the paragraph above had Jewish origins—Schoenberg became a Protestant but reclaimed his Jewish faith in defiance of Hitler. And when in 1929 the Vienna Circle was officially launched, eight of its fourteen members were Jewish. Some of the others, like Viktor Kraft, were commonly taken to be Jewish. Kraft neatly exemplifies the satirist Leon Hirschfeld's advice to travelers: "Be careful during your stay in Vienna not to be too interesting or original, otherwise you might, behind your back, suddenly be called a Jew."

Looking back, many Jewish intellectuals saw the period of the Habsburg Empire as a golden age: the empire's official tolerance and its rich mixture of nationalities and cultures produced a constitutional ambiguity within which Jews, whether traditionalists from Galicia or acculturated Viennese, could find a home. It even provided a paradoxical argument for empire as the most progressive form of government, giving a secure framework of liberal administration to the mutually enriching coexistence of a carnival of voices.

In the 1850s, at the same time as Ludwig's paternal grandfather, Hermann Christian Wittgenstein, arrived in Vienna from Leipzig and began dealing in property, a ditty celebrated the multi-ethnic freedom of the most cosmopolitan city in Europe:

> The Christian, the Turk, the heathen and Jew
> Have dwelt here in ages old and new
> Harmoniously and without any strife,
> For everyone's entitled to his own life.

Before the First World War, Vienna had seen an explosion in its Jewish population, which grew from 2 percent in 1857 to 9 percent in 1900 and then more slowly up to the outbreak of war. It had the third largest Jewish community in Europe, after Warsaw and Budapest. But, even so, the figures understate the leading role that Austrian Jewry played in every area of life—except that of the imperial court and government.

In 1913 a British observer, Wickham Steed, the correspondent of *The Times* in Austria and no friend of Jews, observed that "economically, politically and in point of general influence, they—the Jews—are the most significant element in the Monarchy." Even the Christian Social mayor of Vienna in the 1890s, Karl Lueger, who gained power by harnessing anti-Semitism, making it common currency in political discourse, felt moved to say, "I am not an enemy of our Viennese Jews; they are not so bad and we really cannot do without them. . . . The Jews are the only ones who always feel like being active." Between 1910 and 1913, an unemployed and unemployable Hitler was kept alive in Vienna by a Jewish charity for the homeless—one supported by Popper's father—and by the Jewish shopkeepers who bought his pictures.

Because they were originally excluded from the civil service and the higher ranks of the army, it was in education and the intellectual professions that those of Jewish descent, if no longer of Jewish faith, made their way. In the 1880s Jews constituted nearly a third of enrollments in the classical Gymnasium and a fifth of those in the vocationally orientated Realschule. Jewish enrollments in the faculties of medicine stood at just under half, in law a fifth, and in philosophy a sixth. Robert Wistrich, the historian of the Jews of imperial Vienna, has captured the surge of civic energy that Jewish emancipation released:

With the enactment of the Statute of 1867, designed to grant equal civil and political rights to all Austrian citizens, the Jews showed themselves eager to apply their creative talents.... Jews were among the promoters of charitable institutions; they were the founders of newspapers and educational periodicals, and were prominent in music and literature, economics and politics. As bankers, philanthropists, professors, doctors, writers, and scientists they played their part in the development of Austria.... Moreover, they fully shared with their Austrian compatriots in the defence of the country and participated in many spiritual battles.

They owed the opportunity for this preeminence, and in consequence gave their loyalty, to Austria. So much so that a claim made in 1883 by the chief rabbi of Vienna, Adolf Jellinek, comes as little surprise, given the multinational nature of the empire: "Jews were the standard-bearers of the Austrian idea of unity." A poignant though probably apocryphal tale is of a group of Austro-Hungarian Army officers casting earth into the grave of a fellow soldier: each does it in the name of his own nationality—Hungarian, Czech, Slovak, Polish. Only the Jewish officer speaks for Austria.

But that officer's loyalty would have afforded him no protection from Austria's systemic anti-Semitism. The fatal contradictions are caught in a remark made by the Emperor Franz Josef to his daughter Marie Valerie: "Of course we must do what we can to protect the Jews, but who really is not an anti-Semite?" (This is not too dissimilar from a remark by the former diplomat, biographer, and critic Harold Nicolson: "Though I loathe anti-Semitism, I do dislike Jews.") No matter how comfortable Jewish intellectuals may have felt, at other levels the city was deeply anti-

Semitic. Half a century before Hitler came to power, supporters of Karl Lueger were singing, "Lueger will live and the Jews will croak." For Austrian Jewry, the bitter fruit of continuing success was a deepening of anti-Semitism. The historian Peter Pulzer's verdict is: "If any city in the world can claim to be the cradle of modern political anti-Semitism it is Vienna." And neither Popper nor Wittgenstein would entirely escape its evil.

THE SETTLED EXISTENCE of the German-speaking Jews was threatened by repressive government elsewhere in Europe. Impoverished eastern Jews fleeing from the tsar's pogroms turned up in Vienna as beggars, peddlers, and small tradesmen—so-called *Luftmenschen*, working in the open air, door to door with carts and packs. Living in the poorer parts of the city and with their highly noticeable Yiddish sidelocks, fur hats, and caftans, they seemed a race apart from their middle-class (perhaps now former) co-religionists absorbed in the world of the newspaper office, the lawyer's sanctum, the doctor's consulting room, and the gossip after work in the coffeehouse—and a universe apart from such leading families as the Wittgensteins. Popper's oldest friend, the Vienna-born art historian Ernst Gombrich, spoke of the reaction to these new arrivals:

> If the truth is to be told, Western Jews despised and cruelly ridiculed the Eastern Jews for their frequent failure to understand, adopt and assimilate the traditions of Western culture. . . . I do not feel called upon to judge, condemn or condone this antagonism, but it is a fact that most of the assimilated Jews of Vienna felt that they had more in common with their gentile compatriots than they had with the new arrivals from the East.

Middle-class Jews drew a sharp distinction between themselves as Jews who wore ties, *Krawattenjuden*, and Jews from the East, or *Kaftanjuden.*

As the old century ended, with the Austrian economy suffering severe problems, anti-Semitism became more clamorous. Science took the place of superstition in giving visceral hatred a voice. In the words of the historian Steven Beller,

> [the] success of biology, with its inspiration of social Darwinism, integral nationalism and racialism, threatened the liberal, Enlightenment-grounded assumptions behind Jewish integration in Central Europe. When combined in Vienna with the ability of the Governing Mayor, Karl Lueger, and his Christian Social cronies to harness the really not very modern resentment by the "little man" of Jewish success, this "biological turn" in the form of "scientific" anti-Semitism, effectively destroyed the emancipatory assumptions of Jews (and their allies).

Step outside the major areas of Jewish life and the shout of "*Saujud!*"—"Foul Jew!"—would soon be heard. Theodor Herzl abandoned his dream of assimilation, in which the Jews of Vienna would walk en masse to the Danube and be baptized, and turned to Zionism. In 1897 he was the moving spirit in the creation of the World Zionist Organization. The trumped-up charges of spying for Germany brought against a Jewish French Army officer—the Dreyfus case, which he observed in Paris—were a turning point.

The end of the First World War brought the defeat and dismemberment of Austria-Hungary and a watershed for the Jewish community. The foundation of the short-lived Austrian Republic shattered the imperial consensus within which Jewry had prospered. As the young Popper became a student, anti-Semitism became still more open and vicious. Superficially, Vienna remained its glittering, cultured, cosmopolitan, *fin d'empire* self. But its politics were pregnant with hatred. Although the city was controlled by socialists, many of whose leaders were Jewish, the country was governed by an alliance of Catholic, Christian-Social, pan-German parties in which anti-Semitism was rife.

The war had swelled the city's Jewish population by a third; again refugees came from the East—again strangers to the middle-class assimilationist ideal. But Austria and Vienna were undergoing a painful rebirth. The eminent Hebrew scholar N. H. Tur-Sinai, who fled from Vienna, experienced the change in atmosphere:

> The war had changed the *raison d'être* of the city and of the local Jewish community, and so not only did the newcomers pose a serious and stubborn problem but all Vienna's Jews turned into refugees in a sense. . . . And now the political basis of the Jews had been destroyed. There was no need for Austrians; there were only Germans now.

The implosion of the empire of many nationalities and its recreation as separate nation states ripped away the cloak of cultural invisibility, leaving Jews exposed in what had suddenly become a "German" country. Robert Wistrich succinctly summarizes the impending catastrophe: "With the demise of [Emperor Franz Josef] the floodgates of barbarism would be thrown wide open."

The response of Jewish intellectuals varied. Some emigrated; some joined the socialist or Communist underground; some became interested in Zionism and rediscovered their Jewishness.

Many were so confident of their place in Viennese society that they could not admit to personal danger. Some even supported the government's Catholic conservatism—"better the devil you know." And mixed up with that was, in Malachi Hacohen's phrase, a "hankering back to the 'Austrian idea' of a perfectly liberal and pluralistic state, which had never existed except in the minds of many Jews and some Josephine bureaucrats." The Wittgensteins fell into the "no one will touch us" category. For his part, Karl Popper resolved that the position at home was impossible and that a move abroad was the only option. He took the "Austrian idea" into exile with him, and it informed his vision of the model society. Hacohen believes that "he remained an assimilated, progressive Jew to the end of his life." This was not a description Popper would have accepted. To describe him as a Jew was to court the heartfelt force of his dissent. Nevertheless, it was his Jewishness that drove him out of a career in Austria and into academic exile, from which he returned with much to prove to his peers and little time to prove it. H3 was to be an early showcase.

———— ◆◆◆ ————

10

Popper Reads *Mein Kampf*

Protestant, namely evangelical but of Jewish origin.
— POPPER

IN *Unended Quest*, Karl Popper wrote, "After much thought my father had decided that living in an overwhelmingly Christian society imposed the obligation to give as little offence as possible — to become assimilated." Karl's father, Simon, came from Bohemia, and Karl's maternal grandparents from Silesia (now part of Poland) and Hungary. Jews in these areas were among the most Germanized of the Empire's Jewish subjects. Malachi Hacohen describes their absorption into the prevailing culture once in Vienna: "They sent their children to German educational institutions, moved into white-collar clerical positions, and transformed Vienna's professional elite." Popper's father exemplified this trend, becoming a partner in the legal practice of the last liberal mayor of Vienna, Raimund Grübl (hence Karl *Raimund* Popper). Popper's mother, Jenny Schiff, sprang from Vienna's Jewish *haute bourgeoisie*. Hacohen sees them as forming a household that embodied the virtues of "*Besitz* (property), *Recht* (law), and *Kultur* (culture) that were held in the highest esteem by Viennese liberals."

The decision by Popper's parents to embrace Protestantism rather than Catholicism was also the choice made by most converted Jews: perhaps the Protestant work ethic and the stress on individual conscience made for a more comfortable new home; perhaps to embrace the ruling religion, Catholicism, was a betrayal too far.

What of Popper's own relationship with his Jewish forebears? In his application to the Academic Assistance Council in England for help in leaving Austria in 1936, he described himself as "Protestant, namely evangelical but of Jewish origin." Against the question whether he was willing to have religious communities approached on his behalf, he wrote opposite the Jewish Orthodox section "NO," very firmly. To make his position even more clear, he underlined the word twice.

But being Jewish has been rightly described as belonging to a club from which there is no resignation. Whatever his own feelings, Karl Popper could never escape the interest of others, Jewish and non-Jewish, in his origins. For instance, 1969 brought an inquiry from the then editor of the *Jewish Year Book* as to whether, as he was of Jewish descent, Professor Sir Karl Popper would like to be in the "Who's Who" section, "which includes Jews of distinction in all walks of life." To this Popper replied that he was of Jewish descent but the son of parents baptized years before he was born; that he was baptized at birth and was brought up as a Protestant. And he continued:

> I do not believe in race; I abhor any form of racialism or nationalism; and I never belonged to the Jewish faith. Thus I do not see on what grounds I could possibly consider myself as a Jew. I do sympathize with minorities; but although this has made me stress my Jewish origin, I do not consider myself a Jew.

Nevertheless, he was always conscious of his Jewishness. In 1984, commenting harshly on Israeli policy toward Arabs, he declared, "It makes me ashamed in my origin" *(sic)*. The notion of a chosen people was "evil."

Popper thought that Jews could not hope to remain Jews and be recognized as Germans, and he defended his father's decision to convert:

> This meant giving offence to organized Judaism. It also meant being denounced as a coward, as a man who feared anti-Semitism. But the answer was that anti-Semitism was an evil, to be feared by Jews and non-Jews alike, and that it was the task of all people of Jewish origin to do their best not to provoke it: moreover many Jews did merge with the population. Assimilation worked. Admittedly, it is understandable that people who were despised for their racial origin should react by saying that they were proud of it. But racial pride is not only stupid but wrong, even if provoked by racial hatred. All nationalism or racialism is evil, and Jewish nationalism is no exception.

Jews had to shoulder their portion of the blame for anti-Semitism and for remaining outside mainstream society. It was an approach

Popper as Viennese secondary-school teacher in the early 1930s. He had been trained in the radical approach of letting children think for themselves. Previously they had only learned by rote.

that echoed Karl Kraus: Jews should move out of their self-imposed cultural and social ghetto; by doing so they would achieve deliverance.

In reality, full assimilation was as much a dream as Herzl's mass baptism. And Popper had an alternative vision, inspired by his regard for the empire over which Franz Josef had presided. This, Popper insisted, provided the blueprint for a liberal cosmopolitan society in which diversity could flourish. The Austro-Hungarian Army stood as an apparent illustration of this liberalism, its soldiers having spoken some ten languages. The historical truth was more nuanced: Franz Josef's imperial rule was challenged by the rise of local ethnic nationalisms that he tried to suppress but could not. Such nationalisms were by their nature exclusive—"strangers" were not welcome.

Following the settlement for Europe's central and southern states after the First World War, violent nationalism had free reign. Popper came to see that, as a Jew in the eyes of others, he was in personal danger. Although Hitler did not come to power in Germany until 1933, or absorb Austria until 1938, Popper's assessment of the problems being created for Jewry in Central Europe led him early to grave predictions: "I expected, from 1929 on, the rise of Hitler; I expected the annexation, in some form or other, of Austria by Hitler; and I expected the War against the West." This was farsighted indeed. He had read *Mein Kampf* and taken it seriously. While Popper became a secondary-school teacher and worked to finish what became *Logik der Forschung (The Logic of Scientific Discovery)*, the streets of Vienna were being taken over by "groups of young people, many wearing Nazi Swastikas. [They] marched along the sidewalks singing Nazi songs." In an anecdote reminiscent of Goering's notorious remark about reaching for his gun when he heard the word "culture," Popper recalled an incident just before Hitler came to power in Germany. He met a young man from Carinthia dressed in a Nazi uniform and carrying a pistol. "He said to me, 'What, you want to argue? I don't argue, I shoot.'" Popper thought that this might have planted the seed of *The Open Society*.

In the 1930s, pressures on Austria's Jewish community became ever more acute. Hitler was in power across the border. At home, the clerical-corporatist state presided over mounting discrimination. As Robert Wistrich observed:

> whether Jews belonged to the rich class or the poor, whether they dwelt in the ghetto or appeared on the stage of the Burgtheater, remained committed to their Jewishness or showed off their assimila-

tion—whatever their individual position—all of them formed permanent objects of criticism for the Viennese anti-Semites. The choice of any career by a Jew met with prejudice and hostility.

The Austrian Nazi Party seized control over life in the universities; Nazi students used violence to make them no-go areas for Jews.

Worse was just beyond the horizon. But, by the time it came into view, Karl Popper would be far away from Europe. The door to an academic career in Austria was locked and the Nazis had the key—just at the time when he had hopes of putting schoolteaching behind him. The deteriorating atmosphere led to a decision that was to set his career onto a new path and reinforce that feeling of exclusion from regular academic life that was a permanent part of his outlook. The resentment generated by this sense of marginalization would spill into the Moral Science Club meeting of 25 October 1946.

—————————◆◆◆—————————

11

Some Jew!

If I have exhausted the justification, I have reached bedrock and my spade is turned. Then I am inclined to say, "This is simply what I do."

—WITTGENSTEIN

ONE CHARGE CAN BE leveled both at Popper and, far more validly, at Wittgenstein: of Jewish self-hatred, even anti-Semitism, in their writings.

While Popper was involved with the externalities, the social and political world and the place of Jewish people there, Wittgenstein was predictably focused on the internalities—his own and those of other individuals. He was concerned by the idea of Jewishness as a controlling mechanism for thought. The notion that Jews inherently think in a certain manner was bound into his constant self-torment, and he describes "Jewishness" (integrally part of him) as a limiting or distorting mechanism.

His dawning recognition of his Jewish descent in the 1930s is not easy to fathom, for his family had done their best to put their Jewishness behind them. His paternal great-grandfather was originally called Moses Maier, but in 1808 the family took the name Wittgenstein—after their local princely family in Hesse, the Sayn-Wittgensteins, for whom Moses Maier was estate manager. Many wrongly assumed that Ludwig was a scion of that princely house. *The Times* recorded in his obituary that he came from a well-known Austrian family: "his ancestors included the Prince Wittgenstein who fought against Napoleon."

Ludwig's paternal grandparents converted to Protestantism. The Jewish side of his mother's family had long been converted to Christianity and had heavily intermarried with Christian families; she was a Roman Catholic, and Ludwig was baptized into her faith. In orthodox Jewish terms, as Wittgenstein's maternal grandmother, Marie Stallner, was not Jewish by descent, nor was he—not that this would have kept him out of Nazi clutches, as we will see. Reflecting on Ludwig's family background and his baptism, Fania Pascal, his Russian teacher in Cambridge, did not place him as a Jew. As a Jewish child in the Ukraine, she had experienced the full force of Slav anti-Semitism, and commented that her grandmother would have said of Ludwig, "Some Jew!"

What Ludwig, his brother and sisters made of their Jewish heritage is open to a number of interpretations. A starting point might be the story of the teenage Ludwig and his brother Paul wanting to join a Viennese athletic club that was "restricted." Ludwig thought they might pull off entry with a white lie; Paul thought not. They found another club. If that story is true, however, what is one to make of the fact that, soon after the German takeover of Austria, the *Anschluss*, Paul, "pale with horror," announced to his sisters, "We count as Jews." His horror was justified. In Germany, the Nuremberg Laws were already three years old: they had deprived those classified as Jews of their rights as citizens (while leaving them as German nationals). In effect, the laws made it impossible for Jewish concert pianists to perform in public. Paul must have been acquainted with some of those barred. Vienna and Prague were full of German Jewish musicians seeking work. But his apparent surprise seems curious, bearing in mind his realism over the athletics club.

Another anecdote tells how one of Ludwig's aunts, Milly, asked her brother, his Uncle Louis, "if the rumors she had heard about their Jewish origins were true. '*Pur sang* [full-blooded], Milly,' he replied. '*Pur sang.*'" Later, Milly's granddaughter's view of the family's Jewishness was to prove of the first importance to them.

Then there is Ludwig himself. Early in the First World War, when he was in uniform as a volunteer, he recorded gloomily, "We can and will lose, if not in this year, then next year. The thought that our race will be beaten depresses me terribly, for I am entirely German."

These stories point to the integration of the Wittgensteins into Catholic-Christian Viennese society to a depth where, although they were aware of their Jewish background, it played no part at all in their lives. If they did not actively deny it—though at one point Ludwig felt guilt at having come close to that—it had become invisible even to themselves.

This is not to criticize them. Paul Engelmann, who was Jewish, believed that Wittgenstein was more or less oblivious of his Jewish antecedents until 1938: "In some cases, such as those of

Otto Weininger and Karl Kraus, whom Wittgenstein admired, it is possible to discern the influence of a specifically Jewish environment and certainly they were conscious of it. But Wittgenstein's own ancestry seems to have been too remote to affect him in this way and was more or less forgotten until the *Anschluss.*"

But whatever impression he gave to Engelmann, who had known him since the First World War, Ludwig himself experienced a significant unfolding of his Jewishness in the 1930s. It was during this period that he both wrote his reflections on Jewishness and drew up "confessions" of his sins which in 1931 and 1937 he then read to selected, startled, and often unwilling friends and acquaintances. One of his "sins" was that he had allowed people to believe that he was but a quarter Jewish by descent, rather than three-quarters. Taken at their face value, if said by anyone else (T. S. Eliot, for instance), the following reflections would be condemned as straightforwardly anti-Semitic:

> It has sometimes been said that the Jews' secretive and cunning nature is the result of their long persecution. That is certainly untrue; on the other hand it is certain that they continue to exist despite this persecution only because they have an inclination towards such secretiveness.
>
> Within the history of the peoples of Europe the history of the Jews is not treated as their intervention in European affairs would actually merit, because within this history they are experienced as a sort of disease, and anomaly, and no one wants to put a disease on the same level as normal life. . . . We may say: people can only regard this tumour as a natural part of their body if their whole feeling for the body changes (if the whole national feeling for the body changes). Otherwise the best they can do is put up with it. You can expect an individual man to display this sort of tolerance,

or else to disregard such things, but you cannot expect this of a nation, because it is precisely not disregarding such things that makes it a nation.*

Wittgenstein also indicts himself as capable of thinking only "reproductively," of only taking up the original thoughts of others (non-Jews). In his view this was a Jewish characteristic: "even the greatest of Jewish thinkers is no more than talented (myself for instance)." Yet again he is found generalizing about the Jewish mind. Similarly, in a conversation about religious feelings with a Cambridge friend, Maurice O'Connor Drury, he later described himself as having "100 per cent Hebraic" thoughts.

As Wittgenstein was contemplating what it means to be a Jew, German newspapers and airwaves were drenched in Hitler's campaigning oratory. And, quoting the passages above, Wittgenstein's biographer Ray Monk is moved to observe with what seems great discomfort, "What is most shocking about Wittgenstein's remarks on Jewishness is his use of language—indeed, the slogans—of racial anti-Semitism. . . . Many of Hitler's most outrageous suggestions . . . this whole litany of lamentable nonsense finds a parallel in Wittgenstein's remarks of 1931." Wittgenstein's litany of nonsense included the characterization of Jews as dangerous foreign bodies in the nation's bloodstream. For Wittgenstein apparently—contrary to the views of Kraus and Popper—there could be no assimilation for the Jew and only danger for the host culture in

* This passage is from *Culture and Value*, edited by G. H. von Wright and translated from the German original by Peter Winch, who renders "*Beule*" as "tumour." The director of the Wittgenstein Archive in Cambridge, Michael Nedo, himself a German, points out that "*Beule*" should be translated merely as "bump."

an attempt by Jews to assimilate: precisely the thinking behind the
Nazis' Nuremberg Laws.

Monk, however, distances Wittgenstein from *Mein Kampf*:
Wittgenstein's Nazi language was a "kind of metaphor for himself" as he strove for a new beginning. Between Wittgenstein's two
confessions he had been to the Soviet Union with some idea of
living and working there, either in a university or as a manual la-
borer. And the simplest explanation for his gross remarks about
Jewishness, the trip to the Soviet Union and the confessions is that
they were all part of what Ray Monk identifies as a cleansing
process—the compulsion to dig down to rock bottom and rebuild
from there. Such a process was something that Wittgenstein believed was necessary in politics too, if decline and the old order
were to be rooted out. And that was how he came to empathize
with Stalin's unrelenting drive to remodel the Soviet Union from
the ground up. A remark comes to mind that he made to Fania
Pascal, and that she found so disturbing: that (mental) amputa-
tion had made him healthier. He was like a tree that could be
made healthier only by having all its branches cut off.

There is no sign that Wittgenstein ever regretted what he said
about Jewishness, or that he changed his mind. For the moral that
he drew from his reflections bore no resemblance to the conclu-
sions of *Mein Kampf*, even if the imagery did echo it. Rather, it
was wholly consistent with his answer to the question, "How
should we live?" Jewish characteristics were not to be thought of
as a force for ill. The only culpability for Jews would lie in failing
to recognize their real nature. Honesty required owning up to
one's limitations.

Significantly, these reflections were about being racially Jewish

Adolf Hitler, two days after the Anschluss. *He is "reporting to history" in the Heldenplatz — and to what was said to be the greatest number of Austrians ever to gather in one place.*

rather than about living as a religious Jew. Much later, in 1949, Wittgenstein remarked to O. K. Bouwsma that "he did not understand modern Judaism. He did not see what could be left of it since sacrifice was no longer practised. Prayers and some singing."

WITH JEWISH BACKGROUNDS, both Popper and Wittgenstein were, of course, profoundly affected by the German takeover of Austria on 12 March 1938. Two days later, Hitler stood on the balcony of the Hofburg, the former imperial palace, and was welcomed by hundreds of thousands of ecstatic Viennese — said to be the greatest number of Austrians ever to come together — gathered in the Heldenplatz, the Heroes' Square. He told them, "As Führer and Chancellor of the German nation, I now report to history that my homeland has joined the German Reich."

The *Anschluss* was to bring Wittgenstein face-to-face with the reality of his family's Jewishness, compelling him to deal with high-ranking Nazis in Berlin.

The enforced declaration of assets made by Wittgenstein's sister, Hermine, following the Anschluss of 12 March 1938. The Nazi government demanded this on the basis that the Wittgensteins were Jewish. But in this document Hermine submitted a claim for a change of racial classification.

12

Little Luki

I have just come from the Reichsführer: the Führer has now ordered the physical annihilation of the Jews.
  — SS-OBERGRUPPENFÜHRER REINHARD HEYDRICH

... the nervous strain of the last month or two. (My people in Vienna are in great trouble.)
  — WITTGENSTEIN

IN JUNE 1938, as Karl Popper was still slowly settling down into the mundane frustrations of academic life in New Zealand, Ludwig Wittgenstein was in Berlin, negotiating to save his sisters and other members of the family from the SS.

Although the Nuremberg Laws had been enforced in Germany from 1935, and Austria had been in the throes of pro-Nazi activity, the Wittgensteins had seemed not to feel themselves in any personal danger. Perhaps in their day-to-day lives they were simply unconscious of their Jewish origins. Perhaps they were in denial. Perhaps they were understandably confident of their apparently invulnerable position in Viennese society -- in 1920, hearing of Ludwig's plans to teach in humble village schools, his shaken brother Paul had written to remind him of "the unbelievable fame of our name, whose sole bearer we are in Austria, the immense circle of acquaintances of our father's, Uncle Louis's, Aunt Clara's, the properties we own strewn throughout Austria, our various charities. . . ."

Thinking of the consequences for Germany of a Nazi takeover, Wittgenstein foresaw the worst: "Just think what it must mean, when the government of a country is taken over by a set of gangsters. The dark ages are coming again. I wouldn't be surprised . . . to see such horrors as people being burnt alive as witches." But, in spite of such gloomy prognostications, he seemed unconcerned about the impact on Austria. He plainly did not remember from his time in the high school at Linz how fourteen-year-old Adolf Hitler, two grades below him, had worn a cornflower as a sign of his attachment to a Greater Germany, waved the red, black, and gold flag of the Reich, and shouted *"Heil"* to his friends as the German greeting. So Ludwig denounced as a ridiculous rumor newspaper reports that Germany was poised to send troops into his homeland: "Hitler doesn't want Austria. Austria would be of no use to him at all."

He was a better philosopher than clairvoyant. His opinion was delivered on the very eve of the *Anschluss*. But, when next day Drury told him that Hitler had indeed taken over, Wittgenstein "to my surprise, did not seem unduly disturbed. I asked him if his sisters would be in any danger. [He replied]: 'They are much too respected, no one would dare to touch them.' " It was an echo of brother Paul's observation two decades earlier on the Wittgensteins' position in Austrian society. Privately, however, Wittgenstein was more concerned than he was letting on.

IN VIENNA, the truth dawned quickly—with Paul's shocked acknowledgment that they now counted as Jews. If recognized as such by others, they would have been in the gravest danger. The oppression of the Jews of Austria began at once and was more ferocious than in Germany proper—as if the Austrians were trying to make up for lost time. Within two days of Hitler's Heldenplatz speech, Jewish public officials and judges were ejected from their jobs, senior industrialists were murdered, and doctors and lawyers were forced to scrub anti-*Anschluss* slogans off the pavements with toothbrushes as triumphant crowds jeered at them. Jewish-owned apartments, shops, and businesses were ransacked.

"Nobody was spared," according to a British eyewitness, Norman Bentwich, who recalled "the savagery, the persecution, and the despair with which one of the most cultured Jewish communities in the world, and the third largest in Europe, was stricken. Vast queues gathered outside the consulates of possible host countries. They stretched for miles and were subject to constant attack."

In April, the 99.71 percent vote in favor of union with Germany was not too inaccurate a representation of Austrian feeling once Hitler had made union a fait accompli. Nevertheless, it should be borne in mind that the plebiscite campaign and the poll itself were carried out under omnipresent Nazi pressure, and that the Catholic Church strongly urged its followers to support the *Anschluss* as a matter of "national duty." Soon after the vote, Goering declared that Vienna would be *"judenrein"* (free of Jews) within four years: "They shall go." Hitler's birthplace, Linz, however, was to be cleared of Jews immediately.

At this stage, Nazi policy was to force Jews to emigrate. And the pressure experienced by the Jews is evident from the numbers who left. Between the *Anschluss* in March and the *Kristallnacht* attacks on Jews in November, fifty thousand Jews fled the Ostmark, as Austria was now renamed. By May 1939 more than half of Austrian Jewry had gone.

For the Nazi economy, emigration also meant replenishing the Reich's coffers through despoiling the Jews. To this end, the new authorities moved quickly. Goering ordered the registration of Jewish business: excluding residential property, the value was put at two-and-a-quarter billion Reichsmarks. From 14 April an emigration tax, the *Reichsfluchtsteuer*, was imposed, taking 25 percent of all taxable assets. Once emigration had taken place, the emigrant was classified as an enemy of the Reich and any remaining property valued at over 5,000 Reichsmarks was liable to be seized. From 27 April all capital above 5,000 Reichsmarks had to be registered to prevent its being spirited or stashed away from Nazi eyes.

In November 1938 came *Kristallnacht*, the "revenge" purposefully exacted by the Nazis for the murder of a German diplomat in Paris by a Polish-Jewish youth whose family the German authorities had dumped on the German-Polish border together with 15,000 others of Polish nationality. Throughout "Greater Germany," Jewish shops, industries, synagogues, and communal institutions were destroyed in further violence that the Nazi Party turned on and then, when the leadership felt it had gone far enough, turned off. The damage in Austria was estimated at 4 million dollars. On top of that, Austrian Jews had to bear their part of the fine imposed on the Reich's Jewish community, the *Judenvermögensabgabe*—set at 20 to 25 percent of wealth above 5,000 Reichsmarks. Together, the emigration tax and the fine yielded 2 billion Reichsmarks, poured into armaments.

WHATEVER PAUL'S QUALMS, Ludwig's sisters Hermine and Helene might well have considered themselves to be securely separate from Vienna's Jewish community. They did not participate in its affairs. Family policy was full assimilation—by fiat of Ludwig's paternal grandfather, Hermann Christian, who had forbidden his eleven children to marry Jewish spouses. However, Ludwig's father, Karl, had disobeyed him, marrying a half-Jewish wife—though from a family that had converted to Roman Catholicism. In consequence, Karl's children were partially Jewish by descent, if not at all by outlook. And any sense of invulnerability should have been shattered by the Nuremberg Laws, which were enforced in Austria from 31 May 1938. (Margarete, married to an American, was safe. She spent the war years in New York, while her elder son, Thomas, worked as an agent for the Office of Strategic Services and her younger, John, in Canadian military intelligence.)

The aim of the Nuremberg Laws, Hitler had told the Reichstag at a special session held after his party congress in Nuremberg in September 1935, was to establish a legal regime within which the German *Volk* would be able to establish tolerable relations with the Jewish people. This regime introduced the concept of Reich citizenship but denied it to German Jews. They were made subjects without civic rights, foreigners in their own country: only those of German or related blood could be citizens, *Reichsbürger*, enjoying full political and social rights. In historical terms, the laws could be seen as canceling Jewish emancipation. They also forbade marriage and extramarital relations between Germans and Jews, declaring that purity of German blood was essential for the survival of the German *Volk*. These principles raised a question that was to prove of agonizing interest to the Wittgensteins after the *Anschluss:* who counted as Jewish? Arriving at an answer had delayed the final drafting of Hitler's speech until just before it was delivered.

The key issue was the status of Germans only partially Jewish by descent—what the Nazis termed "*Mischlinge,*" or of mixed race. Infighting between the Nazi Party (wanting to cast the net as widely as possible) and the civil service (for practical reasons wanting to narrow it) was resolved in a series of supplementary decrees. The Nazis were forced to take into account the high degree of assimilation in German society. So much intermarriage across the generations brought the risk of disaffection among the many Germans with Jewish spouses or some Jewish ancestry if the laws were made too rigorous.

For Nazi race theorists, the answer lay in the *Mischling's* grandparents. Those with three fully Jewish grandparents were defined as Jewish. Those with two were Jewish only if they were also Jewish by religion or married to a Jew. However, this would not free any half-Jews from the Nazi terror. They were still non-Aryans, and not full German citizens. Labelled "*Mischlinge* of the first degree," they would face an increasing threat to their existence.

What, then, was the position of Ludwig Wittgenstein and his brother and sisters under these decrees? If their father, Karl, was fully Jewish, giving them two Jewish grandparents, and their mother, Leopoldine, was half Jewish, with one Jewish parent, then, with a total of three Jewish grandparents, they would count as fully Jewish and so cease to be *Reichsbürger*. If their father was not fully Jewish and, say, had only one Jewish parent, that then gave them two Jewish grandparents and they would become *Mischlinge* of the first degree. If they could be shown to have had only one Jewish grandparent, they would be *Mischlinge* of the second degree, and have a still better chance of a decent life, and of avoiding persecution and the loss of their property.

On 15 July 1938 Paul, Hermine, and Helene registered their assets as required of Jews under the new administration but entered a reservation stating that they were seeking a racial reclassification on the grounds that their paternal grandfather, Hermann Christian, was not fully Jewish.

Known as a *"Befreiung,"* a reclassification of Jews to *Mischlinge* of the first or second degree, or of *Mischlinge* to the status of fully German, was possible under a procedure that had existed in the Third Reich since 1935. A *Befreiung* was also possible on so-called merit—service to country or party. Such reclassifications were made by Hitler's deputy, Rudolf Hess, for "mongrels" and their families who had served in the army from the outbreak of war in 1914 or who had fought at the front for Germany or its allies—on the principle that "loyalty should be met with loyalty."

Ludwig and Paul had both volunteered for the front and had been wounded and decorated. So the Wittgensteins' first attempt to escape the clutches of the Nuremberg Laws took the form of Hermine producing a list of Paul's and Ludwig's First World War medals—evidence of the family's courageous attachment to Austria. This category of reclassification was dealt with in Berlin, by the Interior Ministry and the Reich Chancellery, and Hermine and Paul took the medals to "high places" there. But by 1938 the Führer was rebuking those who were forwarding such petitions: "I get buckets and buckets of such applications for exemption, buckets and buckets, *meine Parteigenossen* [my fellow party members]! Obviously you know of more decent Jews than there are Jews in the whole of the German Reich. That's a scandal! I won't tolerate it."

Later in that summer of upheaval of 1938 there was another blow for the sisters. Paul, who was supposed to look after them, decided to emigrate. The case for leaving must have seemed overwhelming. After losing his right arm on the Russian front in 1914, he had painstakingly rebuilt his career as a concert pianist; in his spare time he loved to walk in the countryside. The first would be impossible, the second an invitation to violence for a Jew under the Nazi state. He had marched and given a limb for Austria; now he would not be able to do the two things he loved best. And there was another consideration. Unknown to his family, Paul had two little girls, Elizabeth and Johanna. Their mother, Hilde, was an Austrian Catholic who had studied the piano with him privately, probably on a charitable basis. A talented student, with a love of Beethoven, she came from a typical suburban Viennese family, though not one that would have met the approval of Paul's aristocratic sisters. Hilde's father was a tram conductor. More significant for our understanding of Paul's position are the facts that she was twenty-eight years his junior and was blind. She had lost her sight when ill with diphtheria and measles in 1921, at the age of six. The middle-aged disabled concert pianist and the young blind student were evidently devoted to each

other; theirs is a deeply romantic story. Paul's anxiety for her future can be imagined. He feared that his children would be removed and reared by the Nazi state. His fortune, his family, and
his career all hung in the balance.

Hermine and Helene refused his pleas to go with him. Paul
alone went to Switzerland. From there he journeyed to England,
to tell Ludwig of his family and to consult on where he should set-
tle. His brother advised America. Paul left Europe in April 1939;
safely in New York, he was to prove tough in the Wittgensteins'
subsequent negotiations with the Reich.

Hilde and the two small children became itinerant refugees.
Accompanied by a family friend, they traveled from Vienna to
Italy and, after an anxious wait, to Switzerland, where they remained for some months. Then it was back to Italy to board a
small Genoese liner jammed with refugees, one of the last refugee
boats to escape Italy. On to Venezuela they went, then to Panama,
then Cuba, to be reunited with Paul, and finally to New York.

Paul had fled a month before the Swiss police chief, Heinrich
Rothmund, went to Berlin, proud of his campaign against what
he described as "the Judaization" of Switzerland. In the German
capital he demanded that Jewish refugees should have their passports stamped with a red J, enabling the Swiss frontier police to
identify them and bar their way.

Given the anxieties, it might be asked why all the Wittgensteins did not take advantage of their wealth at home and abroad
to leave. There would have been no problem with emigration permits: at this stage the Nazis regarded the punitive exit taxes on
rich Jews as a means of financing the departure of poorer ones.
However, Vienna was home for both sisters, and, anyway, Helene

could not leave her sick husband, Max Salzer. Now, with Paul gone, the pressure on them growing, and the international scene darkening, the disquiet the sisters felt can be seen in the foolish step they took to protect themselves.

In the autumn of 1938 Hermine and Helene acquired forged Yugoslav passports in the hope that, as Yugoslav citizens, they could leave more easily if the need arose. Almost at once the police moved in on the forgers; the sisters too were arrested. The time they spent in prison was short, but it affected their health. How bleak the future must have seemed to these ladies of retiring disposition and aristocratic demeanor, who had not wanted for luxury or respect and who had made philanthropy a guiding principle for their public life in Vienna. In October 1938 Ludwig told G. E. Moore of the strain he felt over their fate.

The family's only real hope now lay in the reservation that they had put forward in July: they would have to produce evidence showing that their paternal grandfather, Hermann Christian Wittgenstein, was not Jewish, so reducing the number of Jewish grandparents to two and opening the door to a reclassification as half Jewish. The granddaughter of Ludwig's Aunt Milly, Brigitte Zwiauer, had already spearheaded this operation. In September 1938 she had petitioned the *Reichsstelle für Sippenforschung*, the Berlin government bureau for genealogical research, claiming that Hermann Christian was known to be the illegitimate offspring of the princely house of Waldeck and enclosing a photograph of his eleven children on the basis that no one seeing them could possibly think them Jewish. This would have removed one grandparent from the racial equation. Margarete's son John Stonborough thinks it "unlikely but possible" that Hermann Christian

was a bastard; after all, says Major Stonborough, the Meier/
Wittgenstein family seem to have had princely protection when
they lived in Hesse.

However, salvation lay in the Nazi authorities' research into
the family's wealth rather than its ancestry. The Reichsbank in
Berlin began to take an interest in the Wittgenstein fortune,
much of it held abroad in the United States. Hitler's war machine
needed money: in November 1938 Goering told the Reich Defence Council that Germany's reserves of foreign currency had
been exhausted by rearmament. That was despite the foreign currency gained from the *Anschluss* and from the impoverishment of
Austrian Jewry.

How rich were the Wittgensteins? None of his children followed Karl into the steel industry or business generally, and so the
estate might well have been on hold since his death in 1913. It
might then have suffered from the depression and inflation that
affected the new Republic of Austria after the war. However,
Karl's shrewdness in investing abroad—substantially in the
United States, Holland, and Switzerland, after he pulled out of direct involvement in Austrian industry—helped the family survive
better than most the collapse of the economy that so affected the
Poppers. Nevertheless, when Ludwig handed over his inheritance
to his brother and sisters in 1919, Ludwig's eldest sister and head
of the family, Hermine, described the Wittgensteins as having
"lost much of our wealth." Their assets would have taken a further
hammering in the worldwide economic crash of the 1930s. But in
1938 Ludwig told Keynes "my people, who were rich before the
war, are still wealthyish."

These descriptions are, of course, relative—as anyone who sees
Helene's former mansion at Brahmsplatz 4 will realize. A figure

of 200 million dollars in 1920 is put on the family holdings, and even in 1938 the Wittgensteins would have ranked among the richest families in Austria. Brahmsplatz 4 was only one of their houses: there was also Brahmsplatz 7, as well as eleven other city properties, including three major family mansions. Then there was the Hochreith, the vast estate in the country surrounded by acre upon acre of Wittgenstein-owned forest. Paul's registered list of directly held capital assets stretched over five closely typed foolscap pages of international bonds and shares, including holdings in thirty leading American companies. He also declared a collection of antique string instruments—a Stradivarius among them. Small wonder the Reichsbank saw the family's foreign-held fortune as a target and as a bargaining tool for their racial status.

According to one account, the sisters turned for help in making the bargain to a Viennese lawyer specializing in representing commercial interests, Dr. Arthur Seyss-Inquart. He was later to be indicted at Nuremberg as one of the major war criminals and was hanged. By coincidence, when he was detained, the Canadian Army intelligence officer detailed to act as interpreter was Wittgenstein's nephew John Stonborough. Worrying that Seyss-Inquart might recognize his Wittgenstein connection, Stonborough tried to avoid the meeting, telling the American making the arrest, "When he sees the handcuffs, he won't need a translator."

Seyss-Inquart was essentially Hitler's man in Austria, the go-between for the National Socialists in their dealings with Austria's pre-*Anschluss* corporate state. From just before the *Anschluss*, his career was one of irresistible rise: Minister of the Interior in the last days of the Austrian Republic to *SS-Obergruppenführer* and *Reichsstatthalter* (governor) of the province of Ostmark (Austria) in the Third Reich—a job he held until April 1939. Later he would become deputy to the Governor-General of Poland, Hans Frank, and then himself become Governor-General of the Netherlands, where he oversaw the deportation of the Dutch Jews.

In these circumstances, it is hard to imagine that Seyss-Inquart would have acted for the Wittgensteins, although it was with his office, the *Reichsstatthalter's* office, that the family lodged their *Befreiung* claim in July 1938. Nothing became of it. But there was a Seyss-Inquart–Wittgenstein connection—with Seyss-Inquart's brother, Richard, a head of a state institution for problem children and a non-Nazi whom Margarete knew through her charitable work. She sent his family food parcels after the war. It seems likely that it was after Richard's intervention that the authorities approved Paul's departure for Switzerland.

In fact, once the Reichsbank stepped in, the Wittgensteins bargained directly with the authorities in Berlin. Hitler himself took the final decision on their position. The figures show how difficult it was to gain a *Befreiung*. In 1939 there were 2,100 applications for a different racial classification: the Führer allowed only twelve.

Among the supplicants was one whose tragedy puts the Wittgenstein *Befreiung* into context. Harriet Freifrau von Campe was the granddaughter of Bismarck's banker, Gerson Bleichröder—a practicing Jew and at one time the richest man in Germany. Her husband's family was Prussian nobility. After all other avenues to a *Befreiung* had closed, she offered her entire fortune as a donation to the Reich to obtain one, claiming that her real father was not a Bleichröder but "an Aryan." Deportation to a camp in Riga in 1942 was her fate. Her brothers had petitioned for exemption from anti-Jewish measures on the grounds of military service, early support for the Nazi Party, and intention to

Aryanize. *SS-Obersturmführer* Adolf Eichmann, who by December 1942 was dealing with exit permits and evacuation of Jews for the whole German Reich, turned down all the petitions—the brothers were Jews, "especially in the light of the repeated expressions of his will by the Führer." They were exempted from deportation to the East, but fled to Switzerland in poverty.

The happier outcome for the Wittgensteins is perhaps an indication not just of the amount of money on offer but of the complexity the Reichsbank faced in getting its hands on it—it was certainly not a matter for the authorities of the provincial Ostmark, but for Berlin at the highest levels.

Backed by three lawyers—one an American, one responsible for the family holding company, and, significantly, one specialist Viennese lawyer taken on at the suggestion of the Nazi participants in the negotiations—Margarete, Brigitte Zwiauer, and Ludwig dealt with the Reich Chancellery, the Ministry of the Interior, and the Reichsbank foreign-exchange division. The instrument of their racial classification, the *Reichsstelle für Sippenforschung*, seems to have been restricted to taking orders from above.

The basis of the deal was that Brigitte's statement on the family ancestry would be accepted on the transfer of a large portion of the family's foreign currency to the Reichsbank. But, as the threat of war grew, negotiations dragged on. While their family representatives were constantly traveling in search of a settlement—to Zurich, to Berlin, to New York—the sisters lived on their nerves.

Hitler warned that if the Jews "should succeed once more in plunging nations into another world war, the consequence will be the annihilation of the Jewish race in Europe." He divided and took over Czechoslovakia. He made his pact with Stalin. Still the Wittgensteins negotiated with the Reichsbank—and among themselves.

From America, Paul objected to the amount being discussed, and hired a New York lawyer of his own, Samuel R. Wachtell, of the firm Wachtell, Manheim & Group, to look after his interests. Paul's attitude was that he was willing to pay the Third Reich as much as it took to secure his sisters' future, but not a penny more. The Nazis' position was blackmail—and in dealing with crooks you do not show weakness. In a letter to Ludwig, Wachtell claimed that his client had made an offer that was acceptable to the Reichsbank, but that the Reichsbank had found it easy to pressure the sisters in Vienna to persuade Paul to go further. One of the sisters' legal team, a Dr. Schoene, urged Paul to accept the Reichsbank's demands, intimating darkly of the danger ahead for his clients. And there were pleas from Margarete, whom Paul thought was too soft and ready to go too far. He could be very impatient with anyone unwilling to see his point of view.

WHAT WAS LUDWIG'S PART in all this? In the week immediately after the *Anschluss* a friend in Cambridge, the Italian economist Piero Sraffa, had apparently had to warn him not to go to Austria, where he would now be a German national. Wittgenstein acknowledged to himself that becoming a German was a fearful matter—"like red-hot iron"—and that, as a German Jew, if he now went to Austria he would not be able to leave.

On 18 March 1938 he wrote to Keynes, "By the annexation of Austria by Germany I have become a German citizen and, by the German law, a German Jew (as three of my grandparents were baptized only as adults)." It was just as well that these comments about his grandparents, or his earlier "confession" about being three-quarters Jewish, did not reach Eichmann's eyes. Nevertheless, he was still optimistic about the fate of his family, writing, "As my people in Vienna are almost all retiring and very respected people who have always felt and behaved patriotically, it is, on the whole, unlikely that they are at present in any danger."

But Ludwig was now anxious about his own status in Britain, with naturalization on his mind. A fortnight after he became technically a national of the Third Reich, he asked Trinity whether he still had permission to remain in Britain. A. C. Ewing noted that Wittgenstein was keen for his name to appear in the faculty lecture list, as that would facilitate his taking British nationality. Wittgenstein would not have been reassured by a minute of the Faculties Board. This reported that the secretary had been asked by an "alien" to approach the Home Office for permission for him to lecture "at the request of the Faculties Board." "It was agreed that the steps should be taken by the alien and not by the University."

To become British was now imperative. According to Drury, Ludwig was concerned that, in the event of war, he might be interned as an alien. In 1939, after war had been declared, he had a taste of what might have been in store for him when he visited Drury in Pontypridd and was ordered to report at once to the police station. The manageress of the hotel had become suspicious of his foreign name—particularly on hearing Drury joke about the blackout—and had told the police about his arrival.

Wittgenstein had contemplated acquiring British citizenship previously, but had rejected it on the grounds that he did not wish to become "a sham Englishman." Now, with the reality of Nazism being felt across Central Europe, better a sham Englishman than a legal German. He sought Keynes's help in finding a useful solicitor. (Over the years, Keynes's assistance to Wittgenstein was exceptional: Cambridge contacts, cash, a visa for Russia, naturalization.) Early in May 1938 he placed the necessary advertisements in the *Cambridge Daily News* giving notice of his application for naturalization. As it turned out, even with the services of Keynes's recommended solicitor, a Mr. Gwatkin, Wittgenstein did not become a British subject until 12 April 1939, when he took the oath of allegiance. On Friday, 2 June he received his British passport—number 234161. At last he was able to go back to Vienna and thence to the German capital to try to secure his sisters' future.

On Wednesday, 5 July he traveled to Berlin. He put up at a hotel in the fashionable heart of the city near the Potsdamer Platz, the Hotel Esplanade. Opened at the turn of the century, it was listed in Baedeker's travel guide as "of the very highest class"—an accolade otherwise given to only two other Berlin hotels: the Adlon (the Nazi hierarchy's favorite) and the Kaiserhof. He stayed in Berlin for the next day, returning to Vienna on Friday, 7 July. His sister Hermine was proud of how he conducted himself, impressing the head of the Reichsbank foreign-exchange division, probably a Dr. Reinel, with his clarity and grasp of the details. Within a fortnight he was a passenger on the *Queen Mary*, traveling to New York to talk to Paul and his lawyer, Samuel Wachtell. He stayed in a hotel on Lexington Avenue, near the Rockefeller Center, and recalled later that the only person he liked in New York was an Italian shoeshine boy in Central Park, who cleaned his shoes twice. He paid double the asking price.

On 30 August 1939 Helene and Hermine received the piece of

light-blue paper so vital to their destiny. It certified them as being
Mischlinge of the first degree. But this was still an insecure position for them to be in, and for other family members it meant that
they could not pursue a public-service, professional, or academic
career. Ludwig's cousin, Professor Ernst von Brücke, was forced
out of his institute and into exile. However, there was greater relief ahead. On 10 February 1940 a letter was sent from the head of
the *Reichsstelle für Sippenforschung*, Dr. Kurt Meyer, to the Vienna division of the Nazi Party—apparently in reply to a query. It
rehearsed a finding, made without limitation, that Hermann
Wittgenstein, born in Korbach on 12 September 1802, was to be
regarded as of German blood for the purposes of the Nuremberg
Laws. The letter is worth quoting in full:

Re Letter of 12.1.40 Family Mi/Wu

In the family origins case of Wittgenstein and descendants I have
made my decision on the instruction of the Reichminister for the
Interior of 29.8.39, which in turn refers back to an order of the
Führer and Reich Chancellor. In these circumstances the origin
relationships have not been examined by this office in further detail under its own jurisdiction. The decision of the Führer and
Reich Chancellor also applies without restriction to Hermann
Wittgenstein (born in Korbach 12.9.1802) who is to be regarded as
the predecessor of German blood of all the descendants and to
whose grandchildren the legal presumption of section 2(2), second
sentence, of the First Reich Citizenship Order is likewise not
applicable.

Since then origins rulings have been made regarding the nu-
merous descendants of Hermann Wittgenstein, so that their racial
classification for the purposes of the Reich Citizenship Act should

not manifest any further difficulties. If necessary, corresponding origins rulings can be sought in doubtful cases from the Office of Genealogical Research.

Signed Dr. Kurt Meyer

The second sentence of section 2(2) classified those who practiced the Jewish religion as fully Jewish. In other words, if Hermann Christian had ever been a member of Korbach synagogue before his conversion, that was not allowed to stand in the way of his reclassification as of German blood.

The sisters' baby brother, "Little Luki," had helped to pull this off. A few days after Ludwig's New York visit, Paul wrote to his lawyer, Samuel Wachtell, that "a moral claim can never be waived" and approved a settlement. In Zurich on 21 August 1939, Paul signed three memoranda that resolved the family's internal difficulties and made possible the deal with the Reichsbank. "In consideration of his love and affection for his two sisters," he put the remainder of the money and property he left behind when he fled Vienna into a trust fund for Hermine and Helene for a rainy day. He agreed to the liquidation of the family's joint assets held through a Swiss private company—the bulk of their fortune—to find the payment for the Nazis. And he initiated the basis of the agreement that would guarantee *Mischlinge* status and bring security and well-being to the Wittgensteins in Vienna. Their safety had been bought for a sum big enough to interest the Nazi government at the highest levels: a staggering 1.7 tonnes of gold—equivalent to 2 percent of the Austrian gold reserves taken over by Berlin in 1939.

There was a little over a year to go before the deportation of Austrian Jews began in earnest. Wittgenstein's sisters survived without further harassment. But the war drove a permanent wedge between the Wittgenstein siblings. Paul's sisters accused him of putting their lives at risk by being unreasonable, stiff, and unbending in the Reichsbank dealings. Margarete held an additional grudge. She had used her connections to win permission for him to leave Austria for Switzerland. It was given on condition that he return, a condition he promised to fulfill. She believed he had broken that promise. Paul argued that he had observed the letter of his obligation with a fleeting visit. And this sensitive, proud, and private man felt that his sisters had shortsightedly endangered themselves, ignoring his advice to leave while criticizing him for overreacting. They remained divided by anger and mutual incomprehension—the family bonds another victim of Nazism.

The Wittgensteins were not alone in their negotiations with the Reich, which was always anxious to give legal coloring to its expropriations. These were to a limited extent genuine negotiations—though scarcely between equals. The historian Raul Hilberg makes the point that "Aryanization (of property) was perhaps the only phase of the destruction process in which the Jews had some manoeuvrability, some opportunity for playing German against German, and some occasion for delaying tactics. But it was a dangerous game. Time was against the Jews."

MEMBERS OF POPPER'S FAMILY who stayed in Austria were neither as rich nor as fortunate as the Wittgensteins: sixteen of Popper's relatives from his mother's family, the Schiffs, became victims of the Holocaust. His parents were already dead. After Karl left Vienna, his remaining sister—Annie—moved to Switzerland and composed romantic fiction. She had been a dancer before becoming an author. Karl became very angry with anyone who suggested she wrote risqué stories.

Karl Popper applied twice for British citizenship—before the *Anschluss* in 1938 and again in 1941—but first failed the residency qualification and then was caught by the wartime closing of the list. He spent the war years stateless, classified as a friendly alien. On leaving New Zealand to take up his post at the London School of Economics, his alien status provided a series of frustrating hiccups over exit permits and visas to enter Britain. "Our departure problems are appalling," he wrote to Ernst Gombrich. All were finally resolved. The Poppers were among the first to be granted British citizenship when naturalization was reintroduced in 1946. There was one last irritation before finally boarding the MV *New Zealand Star* to leave for Britain and the LSE: "We are not terribly pleased to pay £320 for the pleasure of spending 5 or 6 very rough weeks in the company of strangers. I am particularly concerned about the fact that I cannot endure the smell of cigarettes at sea without getting sick—still, I shall have to get used to it." He finally arrived in Britain at the beginning of January 1946.

Although he had loved Austria, he emphatically turned his back on the past. Asked in 1945 if he would ever consider returning to Vienna, Popper replied, "No, never." After the war, he refused a full-time professorship in Austria, though he broadcast on Austrian and German radio and later, in 1986, for a short time became a visiting professor at the University of Vienna. In 1969 he told the Vienna-born economist Friedrich von Hayek, whom he had first met in London in 1935, that he had considered retiring to Austria. But, he said, he decided against it because of Austrian anti-Semitism. Yet, having made careful inquiries about the legality in Britain of dual nationality, he did eventually retake Austrian citizenship, wanting to make things easier for his wife if he predeceased her. She was not Jewish and still had family in Austria, never losing her attachment to her homeland. Malachi Hacohen sees her as always grieving its loss: "Wherever they went for the next half-century, she was profoundly homesick. She was as much a victim of the Central European catastrophe as he was. But his dreams migrated with him; hers were destroyed."

A moral relevant to both Popper and Wittgenstein could be found in the historian Fritz Stern's plangent epitaph for Gerson Bleichröder, who had riches, influence, and material reward under the Prussian monarchy. "Only the sense of belonging and security, only the sense of safe acceptance had been withheld. And that, perhaps, is the essence of the anguish of assimilation."

But there is an added significance for this story in these events. The two who faced each other in H3 had seen the catastrophe of Nazism and war sweep away their culture, and threaten and destroy their families. But one had access to riches and influence that gave him the freedom to go wherever his inclinations led, personally and philosophically; the other had only himself to rely on in carving out a place in philosophy in which to make his mark, and make his living.

A politically charged murder was to lay bare this gulf of freedom, wealth, social status, and academic acceptance. It would also change the face of the Viennese philosophy in which, to Karl Popper's discomfiture, Ludwig Wittgenstein had played a leading, if distinctly aloof, part.

13

Death in Vienna

Now you damned bastard, there you have it.
—JOHANN NELBÖCK

SHORTLY BEFORE NINE O'CLOCK on the morning of 21
June 1936, Moritz Schlick left his apartment overlooking the
broad formal gardens of the Belvedere Palace at the top of Prinz-
Eugen Strasse, boarded Tram D running gently downhill toward
the center of Vienna, and began his familiar fifteen-minute journey to the University of Vienna, where he held the chair in the
Philosophy of the Inductive Sciences. He got out a few yards from
the stone steps that led up to the imposing main entrance, hurried
through the iron gate and along the cavernous central hall, and
turned right up the stairs toward the law and philosophy rooms.
The fifty-four-year-old professor was already late for his lecture on

the philosophy of the natural world, in which he would examine such topics as causation and determinism and whether men have free will.

Schlick was far from being a scintillating speaker—he delivered his talks in a barely audible monotone—but his lectures were always crowded. Students appreciated the lucidity of his thoughts and the range of his interests, which extended from science to logic to ethics. Silver-haired and waistcoated, he had a dignified and authoritative demeanor, and was popular with the younger generation—celebrated for his kindness and charm. He was also highly influential in academia as the founder and mainspring of the group of philosophers and scientists known as the Vienna Circle, who had made their doctrine of logical positivism the dominant force in philosophy. And, more, he was recognized as the man who had drawn Ludwig Wittgenstein back into philosophy.

As he hastened to his lecture, waiting for him on the stairs that day was an unwelcome figure, a former doctoral student, Johann (or Hans) Nelböck. Nelböck had twice been committed to psychiatric wards for threatening Schlick, and had been diagnosed as a paranoid schizophrenic. In part his obsession with his erstwhile supervisor had to do with a fellow student, Sylvia Borowicka, with whom Nelböck was infatuated. Herself of a somewhat nervous and unstable disposition, she had rebutted all his advances and had compounded what to Nelböck was an incomprehensible error of judgment by expressing romantic thoughts for the Professor of Inductive Sciences. It is not clear whether Schlick—married to an American, and with two children—reciprocated her affection. No matter: in Nelböck's mad imaginings the two were having a torrid affair.

Nor was this the only injury he believed he had suffered at the professor's hands. Following his spells under observation in a clinic, Nelböck had begun a frustrating and largely futile quest for work. His rejection for one job caused a festering wound. A lecturership in philosophy at an adult-education center had been denied to him only when his history of mental illness, which he had tried to keep hidden, was exposed. For this too Nelböck blamed Schlick, the man whose complaints had first landed him in mental care. He brooded over retribution.

Sometimes during lectures—on the analysis of propositions or the nature of truth—when Schlick glanced up from his notes the gaunt, bespectacled figure of Nelböck would be staring back from among the rows of students. Nor was there any respite back in Prinz-Eugen Strasse, where telephone calls would carry insults and menacing threats.

The professor, usually so unruffled, was terrified—he admitted as much to friends and colleagues. He alerted the police and took a bodyguard. But after a time, when the intimidation came to nothing, it was decided to dispense with his protection, and Schlick ceased all contact with the police. "I fear," he told a colleague, "they begin to think that it is I who am mad."

At 9:15, as Schlick reached the half-landing on the stairs to the philosophy rooms, Nelböck pulled out an automatic pistol and fired four times at point-blank range. The fourth bullet, which lodged in Schlick's leg, was superfluous: the third bullet had pierced the colon and stomach, and the first two had punctured the heart. Professor Dr. Moritz Schlick died instantly. Today a brass inscription marks the spot.

Moritz Schlick, founder of the Vienna Circle. After his murder in 1936 by a crazed former student, his enemies condemned Schlick as representative of "a new and sinister strain in philosophy."

THERE WAS A SECOND VICTIM of the shooting. Nelböck had also put an end to the Vienna Circle, already threatened by the increasingly virulent anti-Semitism that pervaded education at all levels in the Austrian Catholic-corporate state. Indeed, in a sad reflection of the scale of bigotry in the city, as the news of Schlick's murder spread, the press willingly assumed that the professor must have been Jewish and his assassin a supporter of the Catholic-corporate government. Dozens of newspaper articles appeared, some viciously attacking Schlick while expressing admiring sympathy for the murderer.

One of these, written under a pseudonym by an academic colleague, "Academicus," sought to put the incident in what the author saw as its appropriate context and to inform readers of the "true facts and motives" behind the killing. The public should understand that Schlick had been a leading representative of a new and sinister strain in philosophy, one hostile to metaphysics and supported by the basest elements in society—Jews, Communists, and Freemasons. Here was a philosophy—logical positivism—that denied the existence of God, denied the existence of spirit, and saw man as merely a cluster of cells. The bullets that killed Moritz Schlick were guided not by the logic of a lunatic but by that of a soul deprived of its meaning of life. Now was the time to wrestle back control of the ideological territory from its pernicious occupying forces:

> Let the Jews have their Jewish philosophers at their Cultural Institute! But the philosophical chairs at the University of Vienna in Christian-German Austria should be held by Christian philosophers! It has been declared on numerous occasions recently that a peaceful solution of the Jewish question in Austria is also in the interest of the Jews themselves, since a violent solution of that question would be unavoidable otherwise. It is to be hoped that the terrible murder at the University of Vienna will quicken efforts to find a truly satisfactory solution to the Jewish question.

A courageous few, including Schlick's son, tried to rebut the chief allegations against the professor. It was not true that he was Jewish, or an atheist. He was a German Protestant; his children had been baptized and confirmed. Nor did he associate with Communists. And nor was it true that he surrounded himself with Jewish assistants. He had employed only one Jewish helper, a librarian by the name of Friedrich Waismann—who had already been sacked in deference to the campaign to rid universities of Jews. It says much about the political atmosphere that no one thought to counterattack on the simple basis that neither Schlick's nor his associates' race ought to be pertinent.

Nelböck was put on trial for murder. And, even in so poisonous a climate, with widespread public feeling that Schlick had got what was coming to him, the court's verdict was a foregone conclusion. Nelböck had been caught red-handed, standing over the corpse, the still-smoking gun in his hands. A witness gave evidence that he had screamed, "Now you damned bastard, there you have it." Anyway, the killer made a willing confession.

His ten-year sentence was on the lenient side—murder was a hanging offense—but the court was mindful that he had confessed and had a history of mental illness. However, because of the gravity of the crime, the guilty man was also condemned to an additional punishment—sleeping on a hard bed, with a new one to be delivered every three months.

In the event, few such backbreaking frames would be required. The Nelböck case quickly became a cause célèbre, and the jailed murderer was transformed in the public eye from a psychologically unstable loner to a pan-Germanic hero. Following the *Anschluss*, he was released on probation and spent the war years doing his bit for the Third Reich as a technician in the geological division of the Mineral Oil Authority. In 1941 his petition for a complete pardon was rejected, and he never regained the title of doctor, stripped from him upon conviction. After all, reasoned officials, if a murderer could be exonerated on the grounds that the strength of his political opposition justified the act, where would that lead?

But by then it had become established in the eyes of the upholders of the corporate state that Schlick had been a Jew-philosopher who had peddled a Jew-philosophy designed to destroy the nobility of the German soul, and that Nelböck, acting out of ideological conviction, had done Austrian philosophy an immense favor. He deserved its gratitude, and in turn that of Austrians and Germans everywhere.

THE MURDER OF Moritz Schlick can be seen as the breaking point of the Viennese nexus between Wittgenstein and Popper. The new philosophy of logical positivism—which held that the point of philosophy was just to clarify the meaning of propositions, and which evolved from scientific method—was now falling prey to the growth of pro-Nazi forces. Genuine debate would have to be suspended, to be exported and reopened in the English-speaking world.

Schlick had arrived in Vienna in more enlightened times. From a family of minor German aristocracy, he had trained as a physicist in Berlin under Max Planck and was personally acquainted with the great scientists of the day. On his appointment to the professorship in Vienna in 1922, it soon became clear that, beyond enhancing the reputation of the university in his own right, he had a rare and unexpected gift: he was a magnet for talent.

Soon he had gathered around him a remarkable group who would meet regularly on Thursday evenings to discuss philosophical issues. They became known as the Vienna Circle, and during the inter-war years they were to overturn centuries-old philosophical assumptions. In particular, they banished ethics and metaphysics from the discipline. Their modus operandi, logical positivism, was for them the wave of the future—which did indeed lash the shores of established philosophy throughout the English-speaking world.

The members included economists, social scientists, mathematicians, logicians, and scientists as well as philosophers—
thinkers of the caliber of Otto Neurath, Herbert Feigl, Rudolf
Carnap, Kurt Gödel, Viktor Kraft, Felix Kaufmann, Phillip Frank,
Hans Hahn, and Hahn's blind, cigar-smoking sister, Olga, an ex-
pert on Boolean algebra. There was also Friedrich Waismann, the
man whose livelihood would become prey to the rise of Nazism
and later to Wittgenstein's brutality.

The Circle also produced a first philosophical link between
Karl Popper and Ludwig Wittgenstein. Wittgenstein was an honorary member and seen as its guiding spirit, though he rejected
both membership and the accolade. Popper never became a
member, though he hoped to, and took on the role of opposition—and so, years before the meeting in H3, of opposition to
Wittgenstein.

A set of academics of disparate temperaments and intellectual
interests, the Vienna Circle might never have gelled into anything resembling a movement had the mild-mannered Schlick
not been such a seductive and good-natured facilitator, quietly
soothing egos and dispelling tension with his gentle humor. It
helped that he was the only one to issue the invitations to attend.
Those who received them felt duly privileged and personally indebted; those who did not, like Popper, felt undervalued.

Technical star of the group was the grand wizard of notation
and symbol, the logician Rudolf Carnap—like Schlick, German-
born. The Circle's political edge came from the economist and
sociologist Otto Neurath, a man of enormous energy and wit, a
lover of life and of women, hard to miss through his worker's cap,
his extravagant and unkempt red beard, and his huge physical
stature—he would sign his letters with a picture of an elephant.

Of the younger crop of academics, the most intellectually pioneering was Kurt Gödel, a thin, bespectacled, and socially awkward man, whose incompleteness theorems were taken to demonstrate that Russell's attempts to derive mathematics from logic were necessarily in vain.

They would meet in a dingy ground-floor reading room in a building on Boltzmanngasse which housed the mathematics and physics institutes. Chairs were arranged in a semicircle in front of the blackboard, and there was a long table at the back for smokers or those who wished to take notes. Rarely more than twenty in number, the Viennese residents would occasionally be joined by visitors from abroad, among them W. V. O. Quine from America, Alfred Tarski from Poland, A. J. Ayer from Britain, and Carl Hempel from Berlin. Like birds feeding off an exotic plant, these out-of-towners then returned to seed their native lands. In this way the Circle's influence quickly spread. In England in 1936, for instance, Ayer published *Language, Truth and Logic*, which transformed him overnight into an academic celebrity. A beautifully brazen polemic, it almost wholly depended for its conception on ideas he had absorbed in the few months he had spent in Austria.

The meetings followed a regular procedure. Schlick would call for silence and read out any letters from his distinguished correspondents (such as Einstein, Russell, the German mathematician David Hilbert, or Niels Bohr) that might bear on a particular point of contention within the group; then the debate would begin on a topic that had been agreed upon the previous week.

Ideologically, what bound them all together was the belief in the importance of applying the scientific method to philosophy — philosophy, they thought, could benefit as much from logical rigor as any other discipline. In this they differed from their peers in what was then the other philosophical capital of the world, Cambridge, who thought it was science which had lessons to learn from philosophy. As Gilbert Ryle put it, "Philosophy was regarded in Vienna as a blood-sucking parasite, in England as a medicinal leech." The real enemy, however, was not Cambridge but German idealism—a tradition that encompassed Fichte, Hegel, and aspects of Kant, and which privileged the roles of mind and spirit over physics and logic. This school, thought the Austrians, exhibited a combination of obfuscation, mumbo jumbo, and muddleheadedness.

There was a fervor about these meetings. The members felt that they were at the center of something fresh and novel; they were slaying some of the most fiery dragons from philosophy's past. And when in 1929 Schlick turned down the opportunity to return to Germany, to a lucrative and prestigious chair (though who would have gladly swapped Vienna for Bonn?), a few of the Circle's fellow members got together to arrange a publication in his honor: a semiofficial manifesto of the Circle's aims and values. It was called *Wissenschaftliche Weltauffassung: Der Wiener Kreis* or *Viewing the World Scientifically: The Vienna Circle*. Three men were named as the intellectual fathers of the movement— Albert Einstein, Ludwig Wittgenstein, and Bertrand Russell.

Einstein was the brightest star in the new scientific enlightenment: his strikingly counterintuitive descriptions of time and space gave the lie—so it was thought—to Kant's claim that there were some things one could discover about the world merely through armchair, head-in-hands, contemplation. One of Kant's examples is "Every event has a cause," which supposedly tells us something concrete about the way the world works, but which is not arrived at through empirical observation. The laws of Newtonian physics were thought to be another example. Yet Einstein had shown up the absurdity of this. For, far from it being possible to deduce Newtonian laws merely by reflection, these "laws" turned out to be false.

Bertrand Russell was the second name on the Vienna Circle's roll of honor. His appeal lay both in his strident advocacy of empiricism—the theory that all our knowledge about the world comes from experience—and in his pioneering application of logic to both mathematics and language. Rudolf Carnap and Hans Hahn were two of the very select band of people who could claim to have consumed and digested the contents of Russell's *Principia Mathematica*, published in 1910–13. Carnap, when he was an impecunious graduate student in Germany during the hyperinflation of the early 1920s, had written to Russell to request a copy of this 1,929-page, three-volume tome, which was unavailable—or unaffordable—and Russell had responded with a thirty-five-page letter detailing all its main proofs. Hahn performed a similar service for the Vienna Circle as a whole, giving them a crash course in Russellian logic and distilling the philosophical essence from the veritable "cemetery of formulae."

But it was for Wittgenstein that the movement reserved its greatest reverence. In February 1933 A. J. Ayer wrote to his friend Isaiah Berlin with his impressions of the group: "Wittgenstein is a deity to them all." Russell, according to Ayer, was seen as merely a "forerunner of the Christ [Wittgenstein]."

In fact by the time Ayer arrived in Vienna from Oxford as a twenty-four-year-old research student, in November 1932, the most intense period of veneration of Wittgenstein had already passed. The German original of *Tractatus Logico-Philosophicus*—

the *Logisch-philosophische Abhandlung*—had quickly caused a stir in the author's home city when it appeared in 1921. Schlick was one of the first to appreciate its originality, and in the mid-1920s the *Tractatus* was read out and discussed sentence by sentence in the Circle—not once, but twice. It was a painstaking process that took the best part of a year.

The process by which Schlick then came face-to-face with the author required equal perseverance. Eagerly anxious for a meeting, Schlick wrote to him in 1924. He was convinced, he explained, of both the importance and the correctness of Wittgenstein's fundamental ideas.

Wittgenstein replied in cordial terms. By this time he was teaching at a village primary school in the lower-Austrian countryside and he invited Schlick to visit him there. Unfortunately, other commitments intervened, and when Schlick finally undertook the trip he discovered that Wittgenstein had resigned and moved on.

It was Ludwig's sister Margarete who eventually brought the two men together. Having given up teaching, her brother had returned to Vienna and was busying himself with the construction of a new house for her in Kundmanngasse. Her son John was a student of Schlick's. In 1927 she contacted Schlick at Ludwig's behest: he would love to meet him, but not with other members of his discussion group, as Schlick had proposed. Schlick's wife recalled her husband leaving the house as though he were off on a pilgrimage. "He returned in an ecstatic state, saying little, and I felt I should not ask questions."

A fellow member of the Circle, Herbert Feigl, wryly declared later that Schlick was so deeply impressed with Wittgenstein's genius "that he attributed to him profound philosophical insights which he had formulated much more lucidly long before he succumbed to Wittgenstein's almost hypnotic spell."

After several such trysts, Wittgenstein finally consented to their being joined by one or two other members of the Circle, including Waismann, Carnap, and, less frequently, Feigl. The venue would vary. Sometimes it would be Schlick's apartment, barely a ten-minute walk from the Palais Wittgenstein in Alleegasse; sometimes they would meet in the Palais itself; sometimes at a Wittgenstein-owned house in between. The only person inconvenienced by these arrangements was the impoverished Friedrich Waismann.

Waismann was sharp and insightful enough to have deserved a job in any university in the world. In Vienna, given the clamor to cut down the number of Jews in academia, the most Schlick could do for him was appoint him librarian, especially as Waismann had not completed his doctorate. From an impoverished family, without money in the bank, holding a poorly paid job, and with a wife and young son to support, Waismann had little option but to live in the densely populated Jewish quarter in the northeast district. His tiny Fruchtgasse apartment was on the Viennese equivalent of the wrong side of the tracks—the crowded if vibrant quarter of Leopoldstadt, across the Danube Canal and outside the Ringstrasse that encircled fashionable and opulent Vienna. Wittgenstein had possibly never set foot in Waismann's part of his home city. And when he was talking about the meaning of intention and gave as an example "I can say, 'Mr. Waismann, go to Fruchtgasse!' What does that mean?" the aristocratic Ludwig might also have been making a social jibe.

Nevertheless, so transfixed was Waismann by the personality of the rich eccentric whose extended family seemed to own half of Vienna that, thin and half-starved as he was, he would dutifully trudge across the city to participate in the gatherings of this inner circle. Indeed, in terms close to those of Feigl's comment on Schlick, the Austrian mathematician Karl Menger—a member of the Circle—described Waismann as having a "grotesque" sub-servience to Wittgenstein, "his idol." "In particular, he changed his opinion whenever Wittgenstein did." Waismann had also become enough of a disciple to pick up Wittgenstein's habit of clapping his hand to his forehead.

Sometimes his schlep, as some fellow residents in Leopold-stadt might have called it, would have been in vain. Often Wittgenstein refused to discuss philosophy and would insist on reciting poetry—his favorite lines at this time came from the work of the Bengali writer Rabindranath Tagore. The crystalline purity and understated spirituality of Tagore's poetry were probably the qualities that Wittgenstein found so attractive. He preferred to read facing the wall. And, as his imprisoned audience of logicians stared at his back, trying hard not to let their impatience show, it might have begun to dawn on them that they had misinterpreted their messiah's message.

> My poet's vanity dies in shame before thy sight.
> O master poet, I have sat down at thy feet.
> Only let me make my life simple and straight,
> Like a flute of reed for thee to fill with music.

To the world of philosophy, one powerful appeal of the Vienna Circle stemmed from their simple, basic tenet that there were only two types of valid statements. There were those that were true or false by virtue of the meaning of their own terms: statements such as "All bachelors are unmarried men," equations such as "2+2=4," and logical inferences such as "All men are mortal; Socrates is a man; therefore Socrates is mortal." And there were those that were empirical and open to verification: "Water boils at 100 degrees Celsius," "The world is flat" (which, being open to verification, is meaningful even if false).

All other statements were, to the Circle, literally meaningless. Thus, since it was impossible to verify whether God existed, religious pronouncements were sent smartly to the intellectual rubbish bin—where metaphysics, too, consequently belonged. In with this "garbage" went pronouncements about aesthetics, ethics, and the meaning of life. Statements such as "Murder is wrong," "One should always be honest," and "Picasso is a superior artist to Monet" could really be understood only as the expression of personal judgments: "I disapprove of murder," "In my opinion people should always tell the truth," "I prefer Picasso to Monet." "Everything is accessible to man," proclaimed the Circle's manifesto. "Man is the measure of all things."

The main function of philosophy, they held, was not to indulge in metaphysics but to sharpen and clarify the concepts employed by the scientist. The scientists were the all-important players on the pitch. The philosopher merely assisted the team by analyzing the tactics of the game. Philosophy would always be subordinate to science.

However, things could not be that simple, even in the Circle's own terms. If statements were deemed meaningful because they were open to verification, what counted as verification? In the Circle's early days, much of its members' energy was taken up with determining that. For instance, how could the maxim "The meaning of a proposition is the method by which it is verified" be adapted to encompass historical propositions such as "William the Conqueror won the Battle of Hastings?" The Vienna Circle believed that science should generate predictions, which could be put to the test. But what verifiable predictions are made by a statement about the Norman Conquest of 1066?

One answer was that the range of tools traditionally at the historian's disposal—archives, correspondence, archaeological evidence, oral testimony, etc.—were the historian's equivalent of the scientist's Bunsen burner, tripod, and test tube, supplying evidence that substantiated one theory rather than another. Moreover, historic propositions did yield predictions, in the sense that, if a proposition were true, one would expect that any related evidence that subsequently turned up would corroborate it.

In years to come, the claim that historical statements gained their meaning only because they were in principle verifiable would strike many people as bizarre. To squeeze all apparently meaningful propositions into this verificationist straitjacket seemed artificial. It meant, for example, weighing propositions about other minds ("Hennie has a headache") solely in terms of the evidence for and against the proposition itself ("Does Hennie request aspirin?"). The alternative, commonsense, view is that a claim such as "Every time the room is emptied of people, the furniture in the room vaporizes (to reappear when they return)" is meaningful: it makes sense, despite being impossible to verify. Even within the Circle there was growing skepticism about the verification principle, which was abandoned almost altogether by the mid-1930s. And later, when A. J. Ayer was asked about the failings of the movement, he would answer, "Well I suppose that the most important of the defects was that nearly all of it was false."

But for a time it was the most fashionable philosophical doctrine in the Western world.

THE THEORY THAT meaningful statements have either to be analytic (where truth or falsity can be assessed by examining the meaning of the words or symbols employed—"all triangles have three sides") or open to observation became known as "logical positivism," and many logical positivists took the *Tractatus* as their Bible. They extracted their principle of verification from the *Tractatus*, and they accepted, as had Russell, one of Wittgenstein's core claims: that all mathematical proofs, no matter how elaborate, and all logical inferences—such as, "If it's raining, it's either raining or it's not raining" or "All men are mortal; Schlick is a man; therefore Schlick is mortal"—are merely tautologies. In other words they give us no information about the actual world; they are devoid of substance: they are only about the internal relationship of the statements or equations. They cannot tell us about Schlick's mortality or whether to leave the house with an umbrella, or whether Schlick is indeed a man.

The total accuracy of the Vienna Circle's interpretation of the *Tractatus* is another matter. Wittgenstein had parceled up propositions into those which can be said and those about which we must remain silent. Scientific propositions fell into the former category, ethical propositions into the latter. But what many in the Circle misunderstood was that Wittgenstein did not believe that the unsayable should be condemned as nonsense. On the contrary, the things we could not talk about were those that really mattered. Wittgenstein had spelt out the point of the *Tractatus* in a letter to a prominent avant-garde editor: "The book's point is an ethical one. . . . My work consists of two parts: the one presented here plus all that I have not written. And it is precisely this second part that is the important one."

A few in the Circle—Otto Neurath among them—came to regard Wittgenstein as a confidence trickster. Rudolf Carnap was particularly struck by the contrast between the Circle's interpretation of Wittgenstein's text and the man himself. The Circle consisted of hard-nosed scientists, dismissive of metaphysics, moralizing, and spirituality—and they initially believed that such rejection was also the message of the *Tractatus*. And yet here, in the flesh, was this poetry-reciting semimystic. As Carnap put it:

> His point of view and his attitude toward people and problems, even theoretical problems, were much more similar to those of a creative artist than to those of a scientist, one might almost say, similar to those of a religious prophet or a seer. . . . When finally, sometimes after a prolonged arduous effort, his answer came forth, his statement stood before us like a newly created piece of art or a divine revelation.

Perhaps inevitably, the misunderstandings and tensions between Wittgenstein and the Circle coterie soon erupted, bringing divisions in their wake. In particular, there was a basic clash of personality with the serene, composed Carnap. Carnap, who believed in the desirability of an ideal language, turned out to be an advocate of the artificial language Esperanto. This innocuous enthusiasm drove Wittgenstein into a rage. Language, he insisted, must be organic.

Although Carnap always deferred to Wittgenstein, his persistent, politely phrased and thoughtful questions about how Wittgenstein reached conclusion Z from assumptions X and Y would be dismissed as the preoccupations of a pedant. "If he doesn't smell it, I can't help him. He just has no nose." The final breach occurred with the publication of Carnap's masterpiece, *Der Logische Aufbau der Welt (The Logical Construction of the World)*. Wittgenstein accused Carnap of plagiarism—a crime that he was always scenting and that he believed was actually compounded in this case by Carnap's acknowledgment in the book of the debt he owed Wittgenstein. Wittgenstein responded, "I don't mind a small boy's stealing my apples, but I do mind his saying that I gave them to him."

But the rupture that smacked of real tragedy—demonstrating Wittgenstein's capacity brutally to dispense with people—came with Waismann, who had been as close to Wittgenstein as any in the Circle. Karl Popper's summary seems justified: "[Wittgenstein] behaved in an inhuman and cruel way to Waismann to whom he was greatly indebted."

Although not the most original of thinkers, Friedrich Waismann had the wonderful capacity of being able to sum up abstruse notions in straightforward and accessible language. For nearly a decade, mostly with cooperation from Wittgenstein, he applied this gift to Wittgenstein's oracular utterances, diligently trying to impose on them a form and a structure. There was even talk, from 1929, of Wittgenstein and Waismann cooperating on a book. Wittgenstein, who never shied away from employing philosophy's most eminent brains as secretarial support, would ask Waismann to take dictation. But in the end the joint-publication plans came to nothing, with Waismann exasperated by Wittgenstein's constant vacillation and possessiveness over his ideas.

In late 1937 Waismann and his family left Vienna as refugees, Popper having recommended Waismann to the British Academic

Assistance Council for help when he himself no longer needed it—although the account Popper gives of this in his autobiography embellishes the truth and is something to which we will return. In any case, Waismann arrived in Cambridge, with his wife and child, having secured a small grant from the Council and a temporary lecturership from the university.

In a strange country, now having to work in an alien tongue, anxious about the fate of his friends and relatives at home, Waismann was in dire need of emotional and professional, not to mention financial, support. It should have been good news that Wittgenstein was the dominant philosopher in the university where he was trying to make a new start.

In fact Wittgenstein was out of the country—in Norway—when Waismann arrived. When he eventually returned to Cambridge, he barely acknowledged the existence of his former collaborator from Vienna. Richard Braithwaite and his wife, Margaret Masterman, stepped in to save the Waismann family from total despair. They provided the refugees with a roof over their heads and some vital extra cash.

The most generous interpretation of Wittgenstein's behavior is that his ideas were now evolving rapidly and he no longer had need of or time for his old Viennese friends. He had been deeply irritated by the publication of the Circle's "manifesto" in 1929, writing to chide Waismann about what he saw as self-satisfied posturing. But that is scarcely adequate justification. Wittgenstein's intense self-concern, his feeling that what people must do is live out whatever role life has given them with maximum honesty—these are the plausible, if not particularly happy, explanations for a failure to put professional antagonisms aside to offer badly needed help. The rebuke Wittgenstein gave when Leavis tipped a boatman whom they had kept waiting also comes to mind: "I always associate the man with the boathouse." Perhaps he always associated Waismann with his poverty and making do in Fruchtgasse.

Academically, Waismann found life intolerable under Wittgenstein's hostile shadow. He was unable to lecture on the topics in which he was most immersed, since these were the areas which Wittgenstein himself was covering in his seminars. With Wittgenstein the more senior in the university, there was no question as to whose interests would prevail. But Wittgenstein even warned students off Waismann. Perhaps he always associated him with being a librarian.

Barely two years later, Waismann moved on to Oxford, where he became Reader in the Philosophy of Mathematics and spent the remainder of his career. He was never a happy exile. Complaining frequently about the absence of coffeehouses, isolated and remote, he had a tendency toward melancholy and depression. Both his wife and his son committed suicide. He did, however, do much to introduce Wittgenstein's new ideas to Oxford, which after the war became the center of Wittgensteinian study. But relations with Wittgenstein himself were never patched up, and the Oxford philosopher Sir Michael Dummett says that when Wittgenstein died in 1951 it was as though Waismann was "released from a tyrant." His lectures, which until then had been almost entirely about Wittgenstein's philosophy, began to explore new ground. Waismann himself died in 1959.

He was far from being the Vienna Circle's only exile—several of the core members were Jewish, and most of the rest had left-wing sympathies. As with so many artists, filmmakers, bankers, scientists, and doctors, Vienna's loss of philosophers was Britain's and America's gain. Carnap went to Princeton via Prague, Feigl to Iowa and then Minnesota, Gödel to Princeton, Menger to the University of Notre Dame, and from Berlin Hempel went via Brussels to Chicago and then New York. Otto Neurath had not returned to Vienna since the right-wing, clerical-corporate Dolfuss Coup in Austria in 1934, when he was traveling in Russia: it was clear that if he, as the most politically active of the group, returned to Austria his life would be in danger. He moved with his wife to Holland, and then, when the Nazis invaded the Low Countries in 1940, he found refuge on a small, crowded boat bound for England, where he died peacefully at the end of the war. Waismann had been one of the last in the Circle to emigrate.

After Schlick's murder, the chair of the Philosophy of the Inductive Sciences was abolished: the appointments committee declared that henceforth the real task of those in the faculty was to teach the history of philosophy. The intent of the Vienna Circle lived on in a dispersed and attenuated form, but elsewhere—in Britain and the United States, not in Vienna.

The Circle's voice can still be heard in a number of philosophical eponyms. In 1931 Gödel published his theorem that scuppered all attempts to construct a logical foundation for mathematics. He showed that a formal arithmetical system could not be demonstrated to be consistent from within itself. His fifteen-page article proved that some mathematics could not be proved— that, whatever axioms were accepted in mathematics, there would always be some truths that could not be validated. Then there is Neurath's Boat. Neurath was an antifoundationalist: he believed that knowledge has no secure substructure. By way of illustration he used a nautical simile: "We are like sailors who have to rebuild their ship on the open sea, without ever being able to dismantle it in dry dock and reconstruct it from the best components."

But it was Hempel's Paradox that went to the heart of the Circle's preoccupation with issues of verification and confirmation. What sort of things would count as confirmation, as evidence, that a theory was true? Hempel's Paradox went like this: suppose you are a bird-watcher and want to assess your theory that all ravens are black. Of course, if you see a white, brown, or green raven then your theory is disproved, falsified. But, equally, surely it is reasonable to take the sighting of black ravens as evidence that your hypothesis is correct. Hempel's insight was that the statement "All ravens are black" is logically equivalent to the statement "All non-black things are non-ravens." Put it another way: if it is the case that all ravens are black, and you spot a green bird, one thing you can say with certainty is "That bird is not a raven." But Hempel then saw that it must be the case that every time you perceive something that is neither black nor a raven, thus confirming the statement that all non-black things are non-ravens, you also confirm the logically equivalent statement that all ravens are black. In other words, you provide evidence confirming this theory every time you see a yellow sun, a white Rolls-Royce, a red robin, a blue bluebell, or a pink panther.

This seems to defy common sense, though trying to work out exactly why is not at all easy. But it shows also that, when Karl Popper began to undermine the Circle's demarcation between those statements which can and those which cannot be verified, he was not as lonely a figure in striking at their positivist project as he was later so eager to tell the world.

14

Popper Circles the Circle

All this made me feel that, to every one of the [Vienna Circle's] main problems, I had better answers — more coherent answers — than they had.

— POPPER

WHAT, THEN, was Karl Popper's relationship to the Vienna Circle?

Popper, like Wittgenstein, never attended its weekly discussions. In Wittgenstein's case this was because he chose not to; in Popper's it was a matter of not being asked. He writes in *Unended Quest* that he would have considered it a great privilege to be invited, but the call never came.

IN 1920, during the meager days after the First World War, a café, the Akazienhof, some three minutes' walk from the University of Vienna's mathematics department, served cheap but wholesome meals on a nonprofit basis to impoverished students. In summer they could eat outside under the shade of the trees. There Karl Popper, then an external (ausserordentlicher) student at the university, ran into Otto Neurath, the most eclectic of the Vienna Circle. This was Popper's first contact with any member of the group; it was Neurath who would eventually describe Popper as its "official opposition."

This title was one in which Popper always gloried. He saw it as epitomizing a characteristic of his life in general and as justifying his philosophical existence. He was not just *an* opponent, he was *the* opponent; not just *the* opponent but the *triumphant* opponent—triumphant not merely over the Vienna Circle, but over Plato, Hegel, and Marx (though he respected both Plato and Marx), over Freud (whom he grouped with astrologers and other pseudoscientists)—and, of course, over Wittgenstein.

Popper was always anxious to finish off what he grandly called the Popper legend. This told that he was a member of the Vienna Circle. Not true, Popper insisted. And it told that, from within the Circle, he had avoided certain philosophical difficulties that had arisen there by changing the verification principle on which a statement was judged as meaningful to one of falsification. Also not true: "The difficulties which beset the Vienna Circle were my making, I invented the difficulties, I showed that their criterion was not practicable, and I didn't try to rescue them out of these difficulties, but I had a completely different problem." His criticisms, he said, soon sowed confusion within the Circle. "But since I am usually quoted as one of them I wish to repeat that although I created this confusion I never participated in it." The stress is on "I" throughout.

Why did Popper always remain outside the Circle's circumference? After all, he did become friends with several of its members, including Carnap, Kaufmann, Kraft, and Feigl, all of whom had a high regard for his abilities. Carnap, Feigl, and Popper even spent a holiday together in the Tyrol in 1932. Feigl said that Popper had "an outstandingly brilliant mind," and Carnap later wrote, "Dr. Popper is an independent thinker of outstanding power."

So Popper had the intellect and the contacts. And he also had the interest in transferring the analytic disciplines of science to philosophy. His first major work, *Logik der Forschung (The Logic of Scientific Discovery)*, published late in 1934, brought the approbation of Einstein and was of equal power to anything that members of the Circle produced. The issue of his exclusion could be put this way: how could the Circle fail to include this young man as he began the work that would bring his international recognition? The answer must be: they could because Moritz Schlick so willed it.

Schlick was not an admirer. His earliest brush with Popper came in 1928, as an examiner for part of Popper's doctoral thesis, which left Schlick unimpressed. But, more importantly, there was Popper's fundamental hostility to Schlick's guru, Wittgenstein — in particular, Popper's attacks on Wittgenstein's rejection of metaphysical propositions and on Wittgenstein's claim that, to be meaningful, propositions have to mirror possible states of affairs. (If the cat is on the mat, then the sentences "The cat is on the mat" and "The cat is on the hat" are both meaningful, since both represent possible states of affairs, even though only one is actually true. Wittgenstein believed that the logical structure of the sentence "The cat is on the hat" reflects the structure of a possible world.) In *Unended Quest*, Popper describes Wittgenstein's long-abandoned picture theory of language—by which language in its structure represents the world—as "hopelessly and indeed outrageously mistaken." A footnote then goes on to criticize Wittgenstein for exaggerating the gulf between the world of describable facts and what is deep and cannot be said: "It is his facile solution of the problem of depth—the thesis 'the deep is the unsayable'—which unites Wittgenstein the positivist and Wittgenstein the mystic."

Popper had been contemptuous of Wittgenstein's philosophy since first encountering it as a young student in the early 1920s. But this disdain only became apparent to a wider audience at a stormy meeting in December 1932—eleven years after the first publication of the *Tractatus*, and when Wittgenstein was already reconsidering the views he had expressed there. It was the decisive moment for Popper's Vienna Circle ambitions, and it took place at what was known as the Gomperz Circle.

Although Schlick's was the most prominent of the groups in the Austrian capital at the time and acquired the widest recognition, there were other, often overlapping, circles. Many intellectuals belonged to several. Heinrich Gomperz, another Viennese philosopher, had a discussion group that focused on the history of ideas. The details of this December gathering, so fateful for Popper, are extremely sketchy. But one account tells how Popper was asked to address the Gomperz Circle and was informed that not only Schlick but also other luminaries of the Vienna Circle, such as Carnap and Viktor Kraft, would be in attendance. There could not have been more at stake for the young teacher. At this stage *Logik der Forschung* had still not seen the light of day, existing only as a vast tome in manuscript form, with the title *"Die beiden Grundprobleme der Erkenntnis theorie"*—"The Two Fundamental Problems of the Theory of Knowledge." This was reincarnated as *Logik der Forschung* after being heavily cut and substantially rewritten. Schlick was the editor of the series in which Popper hoped it would be published, and an impressive showing in front of him might bring the much sought-after call to his Thursday seminar.

Others in the same situation might have pursued a tactic of attentive deference and studied courtesy. But Popper, when tense, was always liable to take an alternative route—no-holds-barred aggression. On that of all nights, he launched into a full-blooded tirade against his philosophical opponents. Wittgenstein was the main target of his derision, being accused by Popper of behaving rather like the Catholic Church in prohibiting discussion of any topic on which he did not have an answer.

Schlick left in disgust halfway through the meeting; later he grumbled to Carnap that Popper had caricatured Wittgenstein. It is a tribute to Schlick's integrity that, despite this contretemps, he subsequently recommended *Logik der Forschung* for publication. But membership of the Circle was something else. If brilliance was one qualification, civility was another. Perhaps a reasonable attitude to Wittgenstein was a third. Popper had effectively flunked his interview. There is no evidence that after the Gomperz evening Schlick ever again considered asking Popper to join his circle. And, according to Joseph Agassi, Popper said many times that the problem he had with the Circle was his refusal to concede that Wittgenstein was a great philosopher.

For the rest of his life Popper would always exaggerate the gap between himself and the Circle. The Circle, he wrote with splendid self-assurance, could be divided into two groups: "those who accepted many or most of my ideas and those who felt that these ideas were dangerous and to be combated."

Yet, beyond the axe grinding, Popper's attack on the Circle's general position was unerringly targeted. He dusted down and polished up a two-hundred-year-old artifact of reasoning to use against the Circle's central tenet.

In the eighteenth century, the Scottish philosopher David Hume first questioned the process of inductive reasoning: just because the sun has risen every day so far, asked Hume, do we have any rational reason for believing that it will rise again tomorrow?

Hume thought not. An appeal to the laws of nature, for example, would simply take us round a circular argument. The only reason we have for believing in the laws of nature is that they have proved dependable in the past. But why should we assume that past reliability is any sort of guide to the future? Bertrand Russell, with his instinct for the arresting image, put the same riddle this way: "The man who has fed the chicken every day throughout its life at last wrings its neck instead, showing that more refined views as to the uniformity of nature would have been useful to the chicken."

Popper showed that Hume's work had important implications for scientific method, where there is a fundamental asymmetry. No number of experiments can prove the validity of a theory (for example, that the sun will always rise), for however often the sun does indeed rise, at some time in the future it may decide to take a well-earned day off. But one negative result can prove a theory false. We cannot logically deduce the validity of the statement "All ravens are black" even if we have tens of thousands of sightings of black ravens and none of any other hue—a blue one might be nesting just around the corner. (A chilling version of this came from the IRA man who pointed out that security for a politician may appear to "work" day after day, but the terrorist has only to be successful once.)

The theory of verification was therefore useless. And, just as fundamentally, the Circle was hoist with its own petard. Its famous slogan that condemned as meaningless all statements that failed its criteria ("meaningful = analytic or verifiable") failed its own test. For the claim that the meaning of a proposition is the method by which it is verified is itself neither true nor false by virtue of the meaning of its terms, nor is it open to verification. The principle cannot be seen, tasted, felt, or smelt; it cannot be experimented on in a laboratory or spotted in the street—so, according to the positivists' own principle, it is meaningless.

Popper entitles one section of *Unended Quest* "Who Killed Logical Positivism?" and feigns remorse in answering his own question: "I fear that I must admit responsibility." He complained, however, that because *Logik der Forschung* was not published in English for another quarter of a century, and because he originated from Vienna and wrestled with many of the same issues as the Vienna Circle, thinkers in the Anglo-American world took him to be a positivist. Neither he nor Wittgenstein could escape the Circle to which they had never belonged.

It was not only outsiders and later observers who associated Popper with the Circle. In place of "verification," Popper had proposed "falsification." A scientific theory could not be proved, but it could be shown to be untrue. For a theory or hypothesis to count as truly scientific, it had to expose itself to disproof. This was interpreted by some in the Vienna Circle as a mere refinement of their principle of verification, a tinkering with their otherwise well-functioning machine. Carnap maintained that Popper exaggerated the differences between his views and the Circle's. Carl Hempel wrote that Popper kept a definite philosophical distance from the Circle—"a distance which I think was excessive; for after all, there was no party doctrine to which the members of the group were committed." And when another member, Viktor Kraft, wrote a short history of the group, he asserted that the Circle's ideas were proselytized in England through, among others, Karl Popper.

Popper always proclaimed that such attitudes represented a serious misreading of his critique. "Verification" had been employed by the Circle to delineate sense from nonsense. But Popper had no interest in drawing such linguistic distinctions. His aim was rather to distinguish science from nonscience or pseudoscience. He certainly did not reject a statement such as "Mahler is a wonderful composer" as gibberish or condemn it as merely subjective: he simply believed it did not fall within the realm of science. "It was clear to me that all these people were looking for a criterion of demarcation not so much between science and pseudo-science as between science and metaphysics. And it was clear to me that my old criterion of demarcation was better than theirs."

It is indisputable, however, that the parameters of Popper's lifetime philosophical interests were established early on, in Vienna. The preeminence he always gave to science and the scientific method—to proof, to logic, to probability—mirrored the foci of investigation in his home city. However far he took the answers, he owed most of the questions to Schlick and his circle, and to Vienna.

So far as the Vienna Circle was concerned, Popper could be

seen as having the last laugh. In 1985 the Austrian government invited him to Vienna to preside over a new institute dedicated to the philosophy of science, the *Ludwig-Boltzmann-Institut für Wissenschaftstheorie*, set up to bring him back to the country of his birth, to Austria's greater postwar glory. It was the final triumph over the Circle.

But in the event, the government's designs on Popper collapsed ingloriously. An official of the Education Ministry told Popper that he would have to submit future work to the government for approval. Penning a raging letter, Popper withdrew. By making him feel forced out by the Austrian government, the offer also brought a curious replay of Popper's earlier life.

WHAT DOES THIS long excursion around the Ringstrasse tell us about the events of 25 October 1946? It explains, of course, how these two Austrians finally came to be face-to-face in a Cambridge college room. But it does more than that.

Popper was personally unknown to Wittgenstein. Nevertheless, their Viennese history points us to the conclusion that, philosophy apart, the aristocrat from the Palais—with his background of English clothes, French furniture, country houses, limitless resources, constant travel, the high acquaintance of cultural giants—instinctively looked down on the bourgeois teacher he confronted in H3. And that he condescended to him with all the insolence of his wealth and position, as he plainly looked down on Waismann but not on the minor aristocrat Schlick.

For Popper, too, Wittgenstein was more than an academic opponent. He was also the Vienna that had always been out of reach even to the son of a respected and socially responsible lawyer. In Wittgenstein he saw the imperial city where riches and status commanded respect and opened doors, the separate territory where inflation-wrought poverty had no place and the Nazis could be bought off. He saw the opposite of the circumstances that had held him back and driven him abroad.

The Ringstrasse was not only the road to H3: it framed their lives.

15

Blowtorch

. . . if we wish our civilization to survive we must break with the habit of deference to great men.

— POPPER

To treat someone well when he does not like you, you need to be not only very good-natured, but very tactful too.

— WITTGENSTEIN

WHATEVER THE SOCIAL AND CULTURAL DIFFERENCES between Wittgenstein and Popper, one similarity of character made it inevitable that H3 would see a raging confrontation: their sheer awfulness to others in discussion and debate.

Physically small and exhaustingly intense, neither man was capable of compromise. Both were bullying, aggressive, intolerant, and self-absorbed—though Wittgenstein once pictured himself more appealingly to Norman Malcolm: "Being timid I don't like clashes, particularly not with people I like."

Popper's strategy in argument has been described by Bryan Magee. Rather than score through identifying minor faults, Popper would carefully strengthen his opponent's case before demolishing its core point. Meeting him for the first time, Magee was struck by "an intellectual aggressiveness such as I had never encountered before. Everything was pursued beyond the limits of acceptable conversation. . . . In practice it meant trying to subjugate people. And there was something angry about the energy and intensity with which he made the attempts. The unremitting fierce tight focus, like a flame, put me in mind of a blowtorch."

While one of Popper's major contributions to scholarship was the insight that for a theory to be scientific it must be open to falsification, he was never happy to accept the application of this principle to his own ideas. It has been said that *The Open Society and Its Enemies* should have been renamed *The Open Society by One of Its Enemies*. Professor John Watkins concedes that Popper was an intellectual bully: "In seminars there were famous cases where someone announced his title, 'What is X?' And Popper would interrupt, ' "What" questions are completely wrong, misguided.' And so the speaker got through his title and nothing more." One such incident occurred in his LSE seminar sometime in 1969, when one of Professor Watkins's doctoral students was due to give an outline of his thesis on "primary and secondary qualities." The young man had hardly started on his exposé before Popper interrupted and began telling him off, implying that he had not at all understood the issue, had no new ideas, and excelled only in subjectivist psychology and the like. "I was not alone in finding Popper's approach a bit rough and unjust."

A recollection from the days spent at the LSE by the author and journalist Bernard Levin—an admirer of Popper—reinforces the point: "One day at a seminar, a fellow student offered an opinion couched in terms greatly lacking in coherence. The sage frowned and said bluntly: 'I don't understand what you are talking about.' My hapless colleague flushed and rephrased his comment. 'Ah,' said the teacher, 'now I understand what you are saying, and I think it's nonsense.' " Stories abound of students, and even members of public audiences, who had the temerity to quote him imprecisely in asking a question being put on the rack until they confessed their error and apologized. "Now we can be friends," Popper would beam. According to Joseph Agassi, "Every lecture course of his started wonderfully and ended miserably because some fool was rude and Popper came down on him, and the atmosphere radically altered from extremely congenial to extremely tense."

Like Wittgenstein, Popper tended to make his students feel useless. Lord Dahrendorf, a German-born sociologist and former director of the LSE, recalls that British students stopped going to Popper's lectures because they were unaccustomed to being treated in this way. And Popper saw no reason not to humiliate even academic colleagues. Ivor Grattan-Guinness, a mathematician, went to his lectures:

> I thought his conduct was awful, frankly. He wasn't encouraging to students, because he knew so much and he laid it on hard. Of course, this made you feel even more stupid than you were to start

with. And the way he used to insult his own staff in front of students like me! There was a nice chap called John Wisdom [a cousin of his Cambridge namesake] who was interested in psychoanalysis. Popper used to insult him in front of the students: "Oh we have somebody here who plays around with Freudianism." I mean, somebody of his eminence speaking this way in front of students!

His assistants were not immune and, like Arne Petersen, who worked with him in the 1970s, could find themselves berated even in public. At the televised opening lecture of Vienna's Ludwig Boltzmann Institute in 1985, Petersen remembers:

Like other foreign members of the Institute present in the seminar I was invited by Popper to pose questions to his opening talk, which had been on *Wahrheitstheorie*, the theory of truth. It so happened that my improvised and somewhat unhelpfully phrased question was understood by him to be a variant of one of the positions he had been attacking in his talk! So there I was, literally in the spotlights, exposed to his devastating argument.

Dahrendorf was amazed by Popper's stamina in dispute: "He would walk up and down and, in his inimitable way, argue and argue and argue and argue. He was marvelously argumentative, and never tired of it." There was a singularity of purpose which would brook no diversion, even on grounds of compassion. In a letter commiserating with Lady Thatcher on her enforced resignation as prime minister, Popper could not forbear from telling her that he disagreed with aspects of her education policy. (The letter remained a draft.)

There was never any question of Popper giving in. The

philosopher Dorothy Emmet had a taste of that in the first of many encounters with him. Popper was in Manchester for a meeting of the Aristotelian Society and she put him up for the night. This came shortly after his arrival from New Zealand and the publication of *The Open Society*. It was a dangerous invitation on her part. In the book, he had not only charged Plato with sowing the seeds of totalitarianism but had also claimed that teaching students *The Republic* turned them into "little fascists." She had written a review in which she defended Plato, commenting that she had been subjected to Plato as a student and that her experience was that reading his works led to a spirit of openness and questioning.

That was not the effect that Plato had produced on Popper. When she introduced herself, he launched into an attack. He broke it off to have dinner with colleagues, then, as soon as he arrived at her house, he started up again "and continued to attack me until, at about midnight, being very tired, I suggested we should retire. Whereupon he completely changed his manner. He said, 'I feel better now I have said all that to you' and he became gentle and rather affectionate. Whenever I met him subsequently he was gentle and affectionate." The trouble was, she told him, that he said things in an extreme way. "Yes I know," he returned, "but I don't really mean it at all." When she discovered that he had only recently arrived in England, she dared offer him some advice: " 'I think you'll find your approach does not work in England. We go in for understatement rather than overstatement.' And he said, 'Do you really mean that? Then perhaps I ought to reconsider my methods.' But he never really did."

Given his propensity to attack with no quarter given, it might come as a surprise that Popper kept any friends—but he did. Apart from the art historian Sir Ernst Gombrich, there was a list that reads like a *Who's Who* of science: Sir John Eccles, Sir Hermann Bondi, Max Perutz, Dr. Peter Mitchell, and Sir Peter Medawar. Four of these were Nobel Prize winners. But the roll of *former* friends is, in comparison, endless, all of them guilty of making an objection to some aspect of Popper's work, no matter how mild or constructive in spirit.

A very few fell out of favor and returned, among them the American philosopher, and author of a controversial biography of Wittgenstein, William Warren Bartley III. A former student turned colleague, he seems to have been something of a son to Popper until, in July 1965, he gave a lecture in which he accused Popper of being dogmatic. He had expected trouble, warning Popper beforehand that he would not like the lecture, and predicting to a member of the audience that Popper would not speak to him afterward. Listening to this lecture, Popper "was completely at a loss." He wrote to Bartley immediately: "I was stunned, bewildered and I hardly knew whether I was dreaming or awake." Nevertheless, in the same letter he suggested that they put the incident behind them, to forget that it ever took place. Even so, their break lasted twelve years—and this time it was not Popper who kept the drawbridge up. The rift was healed when a Californian faith healer told Bartley that he and Popper should be reconciled—and so they were. But the general rule was: once exiled, exiled forever. Even détente was unthinkable. Onlookers were left openmouthed at the ferocity of the rows and the intensity of the rejections.

Among the most notorious examples of both was the breach with the Hungarian-born former Popper disciple Imre Lakatos.

His crime was committed in his contribution to the volume of P. A. Schilpp's *The Library of Living Philosophers* dedicated to Popper: he raised questions about Popper's demarcation between science and nonscience and about his putative solution to the problem of induction—falsification. This was to question Popper's raison d'être. His life was his work: such challenges were impermissible. Those around Popper became accustomed to bitter tirades against Lakatos that continued long after the Hungarian's death. At Fallowfield, Popper's house in the Buckinghamshire countryside, Lakatos and the other philosophers of science who had criticized Popper, such as Paul Feyerabend and John Watkins, were known as "the Wasps' Nest Club."

Another former student turned colleague, Joseph Agassi, made a similar mistake, putting in person his objections to an article Popper had written. The friendship fractured immediately and Agassi became a member of the Wasps' Nest Club. Over the years, Agassi attempted to make peace. But even in his eightieth year Popper could summon up a venomous response—again over a critique of his work:

. . . after the scandalous (because personally aggressive) review you wrote of *Objective Knowledge* (written, according to your introduction to it, unwillingly, and solely because you felt a scholarly obligation to write it); and after a long series of other unprovoked private and public attacks on me (to which I have never responded), I am surprised that you had the courage to write to me those two letters . . . in which you declared that you were well aware that you owed everything to me and denied that you had ever attacked me, not even in that review.

I am an old man, indeed, and still anxious to say some things which I think are important (although I am aware you do not agree). As my time is obviously limited, I do not wish to continue this correspondence.

Popper's supporters say that his academic assaults were aimed at what he saw as pretentiousness—he was intolerant of those who tried to impress. They were never at a personal level—though, for those on the receiving end, separating the academic from the personal might itself seem academic. And it is fair to say that some of Popper's opponents could take the initiative in rudeness. Lakatos, for example, mocked Popper's lectures and advised students to stay away.

Popper's capacity for confrontation, accompanied by sudden bursts of terrifying rage, was not confined to the seminar room—victims could be found in airport terminals and hotels when things were not as expected. These eruptions were then the subject of equally extravagant remorse.

Arne Petersen is forgiving:

I realized that such emotional reactions of Popper were signs of his impatience with mortals, himself included, with our idleness and dogmatism. See how he writes in the autobiography about his disappointment, as a young boy, with the achievements of the contemporary philosophers, his seniors, in whom he had placed such great hopes and who, as it turned out to his dismay, had really not solved what he himself thought could only be elementary problems in philosophy and logic. One can regret the brisk way he did it, but I think he was entirely right in his impatience with what mankind has achieved. Though Popper never brought emotions into his philosophy, they played a great role in his own life, in his decisions and dealings with people. And one should not forget his extremely quick intellect and logical reasoning which made him famous and feared. He was the Socrates of our time.

Not that Popper went in for the Socratic method of instruction through the process of question and answer. Although he loved to be surrounded by students, he preferred to work at home and alone. When he bought his first house in Britain, he is said deliberately to have chosen to live as far from London University as the regulations allowed—thirty miles. His measurement brought him to the village of Penn. There he chose a house at the end of a bumpy road, with the aim of discouraging all but the most determined of visitors. (After his wife's death, he moved to another house in the country, at Kenley in south London, close to the family of his personal assistant, Melitta Mew.)

In Penn, Hennie rigorously excluded all distractions from Karl's work: of course no television, but also after a time no daily newspaper—even though doing the *Times* crossword had been one of her few pleasures. Cooking was excluded too: those who made the trek out there—his assistants and a few friends and coworkers—were rewarded with little more than tea and biscuits. Boiling an egg reportedly caused a great deal of excitement in the Popper household. Students joked that Karl and Hennie were the only people in the world who could convert sugar to protein.

His concentration was phenomenal: John Watkins had an image of him reading a book or manuscript and sucking the meaning out of the content. And his work rate was also prodigious. Weekends meant nothing: he could study, read, and write 365 days a year, pursuing a topic until it was drained dry. The New Zealand story of the writing of *The Open Society* was one of marathon toil. As Hennie typed up version after version, one page became ten, became a hundred, became eight hundred. The effort almost killed them both. "I wrote the book twenty-two times, always trying to clarify and to simplify it, and my wife typed and retyped the whole manuscript five times (on a decrepit old typewriter)." Bryan Magee records that, even in his old age, Popper would frequently labor through the night, ringing him early in the morning, exhausted but exhilarated by his progress. This revealed a staggering dedication; Arne Petersen felt that "what may have started as an act of love became a lifestyle."

Yet coexisting with this workaholic, and with this aggressive, dominating, resentful, vengeful, solitary supremacist, there was an alternative Popper, a Popper who claimed to be the happiest philosopher alive.

Unlike Wittgenstein, this Popper had normal responses to others. He was very empathetic where women were concerned. Wives always knew they could turn to him for help over a troubled marriage. He could be compassion personified, tolerant, even romantic — capable of setting a poem to music for a woman friend. In later life, if someone wrote to him for personal advice, he always took the trouble to reply — often at length. He would give time and consideration to references for his students, repeating them when asked. With his research assistants he got on well, trying to ensure that they received decent pay raises every year from the university and helping them find jobs when they moved on.

Focused on work though he may have been, this Popper had wide-ranging interests and refined musical and literary tastes. In literature his preferences were for the English classics — especially Jane Austen and Anthony Trollope. He read them and reread them, and read them again alongside anyone to whom he introduced the works, to share their pleasure in discovery.

This Popper also enjoyed company, and would bellow with laughter at a broad joke. A favorite concerned a Labor minister by the name of Paling, who called Churchill a "dirty dog." Churchill got up and replied, "The Right Honorable Member will know what dogs do to palings."

This Popper would relinquish austerity given a chance, particularly enjoying Viennese dishes. He relished calf's liver, sautéed potatoes, curd dumplings, apple fritters, a special Austrian sweet pancake called Kaiserschmarrn, chocolate cake. He would have lived on Swiss chocolate. There, perhaps, speaks his deprived late adolescence. But with Hennie there were few chances for such indulgence. She was not interested in food or in socializing. Some understand and forgive his difficult personality as expressing an attachment to a wife whose eternal yearning for Vienna turned into depression, bitterness, and carping, into hypochondria and self-imposed isolation. It is tempting to see the austerity and solitude into which Popper had grown up colluding with the self-deprivation of the wife he adored. His childhood seems to have lacked the warmth of physical affection, setting a model for the adult: he told a friend that his mother had never kissed him — and that he had never kissed his wife on the lips. They slept in separate beds.

He certainly splashed out after Hennie's death in 1985: relaxing more, entertaining more, spending more, living better, plunging into his collection of antiquarian books — the heart of a library worth in all half a million pounds. He acquired in effect, through mutual adoption, a new family when he moved to live near

Melitta Mew, originally from Bavaria, who found him lovable. She also reassured him at long last about his looks. He went on holidays with her, her husband Raymond, and their son, being taken for the grandfather, eating Wiener schnitzel and pistachio ice cream, revisiting the childhood that had been cut short by war and inflation.

16

Poor Little Rich Boy

I told him . . . that imagining him with his philosophically trained mind as an elementary schoolteacher, it was to me as if someone were to use a precision instrument to open crates. Thereupon Ludwig answered with a comparison which silenced me, for he said, "You remind me of someone who is looking through a closed window and cannot explain to himself the strange movements of a passer-by. He doesn't know what kind of storm is raging outside and that this person is perhaps only with great effort keeping himself on his feet." It was then that I understood his state of mind.

— HERMINE WITTGENSTEIN

Towards the end of this tirade [Wittgenstein's] voice had gathered pace and force, and as he uttered the last few words he seemed as if he was administering a coup de grâce to some cowed animal.

— THEODORE REDPATH

WHILE POPPER REMAINS recognizably human despite his aggressive approach to debate and disagreement, there is an unearthly, even alien, quality to Wittgenstein's dealings with others. "[Wittgenstein's] extraordinary directness of approach and the absence of any sort of paraphernalia were the things that unnerved people," was the novelist Iris Murdoch's verdict on him. "With most people you meet them in a framework, and there are certain conventions about how you talk to them and so on. There isn't a naked confrontation of personalities. But Wittgenstein always imposed this confrontation on all his relationships."

Murdoch had little personal contact with Wittgenstein, meeting him only briefly. But she reflected on how deep his influence had been on her all the same, mediated through her novels. Her biographer Peter Conradi points to how Wittgenstein's presence is to be felt in *Under the Net*. A character, Nigel, quotes him in *Bruno's Dream*. "Wittgenstein" is the first word of *Nuns and Soldiers*. The speaker, Guy, goes on to say, "It was his oracular voice. We felt it had to be true." *The Philosopher's Pupil* has this description of the philosopher: "A simple lucidity seemed always close at hand, never achieved. He longed for thoughts which were quiet and at rest. . . . The crystalline truth, not a turgid flood of mucky half-truths."

Murdoch went to Cambridge in October 1947 as a philosophy research student at Newnham, hoping to see something of Wittgenstein, only to find that he had given up the chair of philosophy at the end of the summer. Her principal connection was with his disciples: it must have been what she saw of his influence on them that led this author of moral comedies to believe that there was something demonic in him, to comment that he was an "evil man," and to doubt his moral awareness, saying that he had only "a dream of religion."

Despising professional philosophers, Wittgenstein was in favor of his students abandoning the subject. The aptitude of the student meant nothing to him: he counseled one of his most brilliant pupils, Yorick Smythies, to work with his hands, even though Smythies was so ill-coordinated that he had difficulty in tying his shoelaces. Manual work was good for the brain, Wittgenstein told him. Smythies's parents and those of another of his pupils who went off to work at a factory bench, Francis Skinner, might well have regarded him as a malign genius for persuading their intellectually gifted sons to forsake academia.

What were the roots of this dominion over friends and students? An illuminating insight comes from Wittgenstein's successor as Professor of Philosophy, G. H. von Wright: "No one who came in touch with him could fail to be impressed. Some were repelled. Most were attracted or fascinated. One could say that Wittgenstein avoided making acquaintances but needed and sought friendship. He was an incomparable but demanding friend."

Von Wright has also made clear quite how demanding Wittgenstein's friendship could be, describing a process not unlike brainwashing or, indeed, joining a cult. "Each conversation with Wittgenstein was like living through the day of judgment. It was terrible. Everything had constantly to be dug up anew, questioned and subjected to the tests of truthfulness. This concerned not only philosophy but the whole of life."

The surviving eyewitnesses of the confrontation in room H3 recall the unease and trepidation they felt in their dealings with

Wittgenstein—even close friends such as Peter Geach. Geach remembers long and intellectually taxing walks in the countryside around Cambridge as "work rather than pleasure." Wittgenstein was "brutally intolerant of any remark he considered sloppy or pretentious." Stephen Toulmin attended Wittgenstein's twice-weekly seminars: "For our part, we struck him as intolerably stupid. He would denounce us to our faces as unteachable."

Sir John Vinelott was also at the seminars, and the note he strikes is of being in the presence of a charismatic prophet: "The impression he made upon one was of somebody whose life was consumed with a passion for inquiry, for discovery, for intellectual excavation, and who was profoundly honest and simple in his style of life. He was a difficult man because his honesty and his directness were uncomfortable to most ordinary people." And then there was the sheer physical impression he made: "Very withdrawn, a huge great forehead, very penetrating eyes, but above all, when he concentrated standing up talking to somebody . . . he had so many anxiety lines on his forehead that they made a checkerboard. I've never seen a human face like it in my life before."

Looking at Wittgenstein's relations with others, the most obvious feature is his will to be the dominant, if not the only, voice. Leavis concluded, "Wittgenstein's discussions were discussions carried on by Wittgenstein." This is so well attested that a single instance will suffice. And, having recorded Dorothy Emmet's ordeal with Popper, let us be evenhanded and report her experience of Wittgenstein, whom she met only once, during the Second World War. She had traveled from Manchester, where she was teaching, to give a paper to a branch of the British Institute of Philosophy in Newcastle. Her host, a research biochemist, had fallen into conversation with a hospital orderly who came to collect a piece of equipment and, on discovering this strange Austrian had an interest in philosophy, had invited him along. Emmet duly arrived and was told the news: "She said to me, 'I hope you won't mind if Wittgenstein comes.' I said, 'What!' I gave my paper, and when it came to discussion he brushed it aside and the talk centered on him. I was fascinated by seeing Wittgenstein in action, and so did not mind that my paper was disregarded."

Was it simple arrogance that came across on a first encounter with Wittgenstein? Leavis thought not: Wittgenstein's behavior was rather "a manifestation of the essential quality that one couldn't be very long with him without becoming aware of—the quality of genius: an intensity of concentration that impressed itself on one as disinterestedness." Nevertheless, there was some criticism, in that "Argument once started, he exercised a completeness of command that left other voices little opportunity—unless (which was unlikely) they were prepared to be peremptory, insistent and forceful." Of course, Dr. Leavis had not met Dr. Popper.

Wittgenstein's manner of debate was satirized in 1930 by "An epistle on the subject of the ethical and aesthetic beliefs of Herr Ludwig Wittgenstein (Doctor of Philosophy) to Richard Braithwaite Esq. M.A. (Fellow of King's College)," a poem published in a Cambridge avant-garde magazine, *The Venture*, by a bright young undergraduate, Julian Bell:

> In every company he shouts us down
> And stops our sentence stuttering his own;
> Unceasing argues, harsh, irate, and loud,
> Sure that he's right and of his rightness proud . . .

Bell, who died driving an ambulance in the Spanish Civil War in the summer of 1937, was the son of the artist Vanessa Bell, the nephew of Virginia Woolf, and the grandson of Sir Leslie Stephen. In other words, he was set fair to be an undergraduate at King's, a member of the Bloomsbury set, and an Apostle. It was at meetings of the Apostles that he saw Wittgenstein in action. He thought of proposing a thesis on Wittgenstein to the English board when contemplating a doctorate, but was discouraged by Moore. Politically, he was very much part of the university's left wing, and was for a brief while the lover of Anthony Blunt, the future Soviet spy. Blunt appears to have taken a profound dislike to Wittgenstein and may well have encouraged Bell to write the poem.

The Venture featured names destined for fame and redolent of their age. It was edited by Michael Redgrave and Anthony Blunt, and contained work by a galaxy of future poets, authors, and critics of distinction—including Louis MacNeice, Clemence Dane, Malcolm Lowry, John Lehmann, and William Empson. This edition, the fifth, sold out in three weeks—whether because of Bell's poem cannot be known, though Fania Pascal recalls that when it appeared "the kindest people enjoyed a laugh; it released accumulated tension, resentment, perhaps fear." The "epistle" is long—304 lines—and is worth looking at in some detail for the picture it gives of Wittgenstein not long after he returned to Cambridge, drawn by someone who was definitely not an admirer, but who—though only "a puny Jonah"—has decided that he will

> The great behemoth of the seas defy;
> Whose learning, logic, casuistry's so vast
> He overflows the metaphysic waste.

The author notes the impact of Wittgenstein on

> The rational commonsense, the easy rule,
> That marked for centuries, the Cambridge school.
> But who on any issue saw
> Ludwig refrain from laying down the law?

And there is a know-all aspect to Wittgenstein that grates on the author. The word "omniscience" recurs, as here:

> With privileged omniscience soaring high
> He sees the Universe before him lie;
> Each whirling lost electron's motion planned
> He reads as easy as a watch's hand . . .
> Ludwig's omniscient; well, I would be civil,
> But is he God Almighty, or the Devil?

But at the heart of the poem is another aspect of Wittgenstein's personality. As a haven of tolerant dialogue, where arguments are overturned through lightly handled (if envenomed) irony, Cambridge might not approve of shouting down, claims of omniscience, or unrestrained attacks on established views, but these would not be unknown. Perhaps fellowships and chairs might eventually be withheld as a mark of disapproval. Here, however, is a deeper divide. Wittgenstein is a mystic whose asceticism cuts him off from the common pleasures of life—someone who looks to a secret source for knowledge of the world:

> . . . knowing by his direct experience
> What is beyond all knowledge and all sense.

And the author apostrophizes Braithwaite to declare why they should follow Wittgenstein:

> Ah, Richard, why must we, who know it vain,
> Seek value through this tortured maze of pain;
> When we so easily in matter find
> Every delight of body and of mind.
> I pity Ludwig while I disagree,
> The cause of his opinions all can see,
> In that ascetic life, intent to shun
> The common pleasures known to everyone.

There is that removed quality, outside the everyday social experience. If his life had been one of religious contemplation or charity, "saintly" might have been the impression. But no saint had so brutal a manner toward his fellow men.

In August 1925 J. M. Keynes and his new wife, Lydia Lopokova, were a fortnight into their honeymoon in Sussex when Wittgenstein arrived for a short visit. Keynes's biographer, Robert Skidelsky, tells the tale: "Lydia remarked to Wittgenstein, no doubt brightly, 'What a beautiful tree.' Wittgenstein glared at her: 'What do you mean?' Lydia burst into tears." To add injury to the insult to his bride, Keynes had paid Wittgenstein's fare. But Lydia had not been singled out. Wittgenstein first met Joan Bevan, the wife of his last doctor, just after he had returned from the United States. She observed, "How lucky you are to have been to America," only to have him reply, with force, "What do you mean by lucky?"

Here was no simple lack of manners or unfortunately maladroit style. Wittgenstein was not in the world of polite conversation and social chitchat. Clarity of meaning was all, and he went straight to it—no matter what. When his Russian teacher, Fania Pascal, confessed to a blunder she had made, Ludwig weighed it up and replied simply, "Yes, you lack sagacity." And when he wanted you to do something other than what you were doing, she says, the effect was disturbing. "He conjured up a vision of a better you, undermining your confidence." And this feeling remained to haunt the victim—"What would Wittgenstein say to your doing this, saying that?" Some of those he upset or unsteadied in this way recognized that his thrusts were generated by the difference and unity of his outlook, the virtue of his vice. Nevertheless, Fania Pascal suffered a wound that still hurt years later when he harshly told her that if she gave a Workers' Educational Association course on current events it would be evil and damaging. "The wholeness of his character makes partial criticism of him appear carping, but I could never look on his ability to find out the weak spots of another human being and to hit out as anything but a flaw. The knowledge that he was a man of great purity and innocence cannot alter my feeling."

LIKE POPPER, one of whose sisters committed suicide and whose uncles were not on speaking terms with his father, Wittgenstein came from a troubled family background. His father was impossible to please and could be tyrannical toward his children—his sons particularly, though his daughters did not escape his tongue and his arbitrary decisions. In front of her, he referred to his daughter Helene as "the ugly one." They were scared of but captivated by him.

Three of Wittgenstein's brothers committed suicide: two, Hans and Rudolf, as young men under obdurate pressure from their father to abandon music as a career and follow him into industry.

The third, Kurt, shot himself at the end of the First World War, unable to accept that his men had surrendered after refusing to follow him into battle. On his mother's side, the family had a strong military tradition that made his troops' action insupportable. It must, however, be remembered that this was an age of self-destruction. "Men everywhere are becoming more weary of the burden of life," declared an article in the *Contemporary Review* toward the end of the nineteenth century, while other European journalists claimed that in no other epoch had suicide been as widespread: it had become a kind of epidemic. Social disintegration, the emancipation of the individual, poverty, and the influence of certain philosophers—including Schopenhauer and Kierkegaard—were among the factors blamed.

While poverty was never a problem, Wittgenstein himself was suicidal too—restless, tormented, and obsessed with the sense of his own sinfulness. In 1913 David Pinsent recorded how Wittgenstein had told him "that all his life there had hardly been a day, in which he had not at one time or another thought of suicide as a possibility." In 1919 Wittgenstein wrote to Paul Engelmann, "Just how far I have gone downhill you can see from the fact that I have on several occasions contemplated taking my own life. Not from despair at my badness but for purely external reasons."

Like Popper, Wittgenstein demanded solitude. He was in the habit of retreating to cold and desolate parts of Europe—to the west of Ireland, to Iceland or to Norway, where in 1913 he built himself a wooden house. "He swears he can never do his best except in exile. . . . The great difficulty about his particular kind of work is that—unless he absolutely settles all the foundations of Logic—his work will be of little value to the world. . . . So he is off to Norway in about ten days," Pinsent told his diary.

Wittgenstein did some of his best work in isolation. But he could not switch off the tap of ideas wherever he was. It was said that philosophy came to him, rather than he to philosophy. He could relax only with difficulty. He lost himself in going to the movies to watch musicals and westerns—sitting as close to the screen as he could—and in American hard-boiled detective magazines. But hard-boiled fiction was far from the only literature that he liked. He regularly read Sterne, Dickens, Tolstoy, Dostoevsky, and Gottfried Keller. He admired Agatha Christie, and also P. G. Wodehouse, whose story "Honeysuckle Cottage" he thought exceptionally comical. St. Augustine's *Confessions* was also on his bookshelf, and William James too. He could discuss Kierkegaard and Cardinal Newman, was familiar with Molière, Eliot, and Rilke, and recommended Faraday's *The Chemical History of a Candle* as an illustration of fine popular science. Yet, as Engelmann explained, "He did indeed enjoy reading good detective stories, while he considered it a waste of time to read mediocre philosophical reflections."

The lack of intellectual pretension of the movies and detective stories was presumably what Wittgenstein found so palatable. There is something rather touching in the idea of this most rigorous and demanding of intellects absorbed in the adventures of the Los Angeles private detective Max Latin, a tough-guy crusader against the forces of evil. Latin was the creation of Norbert Davis, a successful but second-division operator in the Hammett/Chandler school of the hard-boiled and one of Wittgenstein's favorites. There is nothing amiss with Latin's moral sensibility—though he strives to conceal it under a heavy cloak of cynicism as he deals with clients and police in the booth of a steamy, packed restaurant that he uses as an office. (He actually owns the restaurant.)

But if necessary—and it often is necessary—he is not afraid to use violence:

> He took one catlike step toward her and hit her. His fist didn't travel more than six inches, and it landed with a sharp smack just below the hinge of her jaw just below her ear. Teresa Mayan whirled around with a graceful rustle of silk, fell across the divan, and rolled off on the floor. She lay motionless, face down. Latin dropped instantly on one knee and one hand, like a football linesman getting ready to charge.

The writing is stripped down to the minimum, like the supremely functional architecture of the house in Kundmanngasse that Wittgenstein joined in building for his sister Margarete, and perhaps this economy was a reason why Norbert Davis and the tough-guy detective genre appealed to him.

A KEY TO COMPREHENDING what drove Wittgenstein is to see him as living a passion for exactitude in all things: a thing was either exact or it was not, and if it was not, it was literally too painful to endure. Leavis wryly describes Wittgenstein searching, unasked, through his gramophone records and finally putting on Schubert's Great C Major Symphony.

> A moment after the music began to sound he lifted the tone-arm, altered the speed and lowered the needle on to the record again. He did this several times until he was satisfied. What was characteristic about the performance (Wittgenstein's) was not merely the aplomb with which he ignored our—my wife's and my—apprehensive presence, but the delicate precision with which he performed the manoeuvre. He was, in fact, truly and finely cultivated,

and as part of his obvious cultivation, very musical; and, having absolute pitch, judged and acted instantaneously on hearing the opening bars.

No question of asking his hosts first: the right pitch had to be found.

That this was not a simple case of the demands of absolute pitch but the expression of something far more deeply rooted is shown by another recollection of Ludwig by his sister Hermine, about his part in the building of the triumphantly modernist house in Kundmanngasse in 1926. The architect for the project was Ludwig's friend Paul Engelmann, a student of Adolf Loos's. The precise part played by Ludwig in the architecture is a matter of debate, but there is no question of his influence on the detail: door and window frames, window catches, right-angled radiators so symmetrical that they could be used as plinths for works of art, ceiling heights. There can also be no question about the outcome: a triumph of total design, rigorously simple and balanced, ever poised and graceful, the essence of harmony. But getting there must have been a nightmare for the manufacturers and builders. "I can still hear the locksmith asking him, in connection with a keyhole, 'Tell me, Herr Ingenieur, does a millimeter here or there really matter so much to you?' Even before he had finished speaking, Ludwig replied with such a loud and forceful 'Yes!' that the man almost jumped with fright."

Happily, money was no object. To obtain the precision that is the secret of their beauty, the radiators and their supporting legs had to be cast abroad—Austrian casting was not up to it. Problems with Wittgenstein over the dimensions of the doors and window frames caused the builder to break down in a fit of sobbing. And then, just as the finished house was ready to be cleaned up, Ludwig "had the ceiling height of one of the rooms, which was almost big enough to be a hall, raised by three centimeters. His instinct was absolutely right and it had to be followed."

The "Herr Ingenieur" also invented the paint for the walls—mixed to give a special surface texture, warm and lustrous, to their off-white color. The doors and window frames were painted in a green so dark it was almost black, while the floors were made of black-green marble. Most of the double windows had blinds between the two panels of glass, but upstairs in the main rooms long transparent white curtains hung from ceiling to floor, allowing the window frames and catches to be seen. The furniture came from Margarete's outstanding collection of period French pieces. In *Culture and Value*, Wittgenstein reflects, "The house I built for Gretl is the product of a decidedly sensitive ear and *good* manners, and expression of a great *understanding* (of a culture, etc.)."

In the same spirit, Wittgenstein changed the proportions of his windows in Whewell's Court with strips of black paper. And, if more evidence is required of his need for exactitude, there is his mixing of medicines at Guy's Hospital when he was a dispensary aide during the Second World War. He had to prepare an ointment, Lassar's paste, for the dermatological department. The sister on the ward said that no one had ever produced Lassar's paste of that quality before. It was the same with even the most casual of tasks. Sitting behind his uncle on a bus, John Stonborough watched him help an old man to put on a rucksack, and marveled at his insistence on every strap being just in the right place.

His eldest sister, Hermine, who had been like a mother to her "little Luki," saw him as possessing a mind that could penetrate to the heart of things, "to grasp in the same way the essential nature of a piece of music or sculpture, a book, a person, even on occasion—however curious it may sound—a woman's dress." When Mrs. Bevan was invited to a reception for King George VI and Queen Elizabeth at Trinity College, Wittgenstein screwed up his face upon seeing her coat, fetched a pair of scissors, and snipped off two buttons. After the operation, she says, it looked much more elegant. Hermine observed that he suffered "almost pathological distress" in any surroundings that were uncongenial to him.

ALL THESE IDIOSYNCRASIES might have been no more than disturbing or upsetting to friends and colleagues, students and visiting lecturers, builders and craftsmen—a price to be paid for being in touch with so profound a thinker. They might all have been put down as childish in their selfishness and lack of social grace. Wittgenstein had a sense of humor, a capacity for playfulness, that does seem to have had precisely a childlike quality to it. A favorite joke went like this: a fledgling leaves the nest to try out its wings. On returning, it discovers that an orange has taken its place. "What are you doing there?" asks the fledgling. "Ma-melaid," replies the orange.

But alongside what Iris Murdoch described as Wittgenstein's capacity for directness of approach and his absence of "paraphernalia," along with the passion for exactness and the childlike playfulness, there was something more. What crops up again and again in the many recollections of Wittgenstein is his power to arouse fear, whether in friend or in foe. Von Wright believed that "most of those who loved him and had his friendship also feared him." Even Joan Bevan, who looked after him in her own home as he was dying of prostate cancer, "was always afraid of him." This was no ordinary anxiety at, say, having an argument torn to shreds. There was fear of violence.

An incident recorded by Norman Malcolm involved a paper given to the Moral Science Club in 1939 by G. E. Moore. In this paper Moore attempted to prove that a person can know that he has a sensation such as pain—a view to which Wittgenstein was adamantly opposed, thinking it meaningless rather than impossible. Wittgenstein was not at the MSC meeting, but on hearing of the paper, according to Malcolm, he "reacted like a war-horse." He went to Moore's home. In the presence of Malcolm, von Wright, and others, Moore reread the paper. "Wittgenstein immediately attacked it. He was more excited than I ever knew him to be in a discussion. He was full of fire and spoke forcefully and rapidly." His performance was not just impressive, it was "frightening."

Malcolm himself was on the receiving end of this power when Wittgenstein visited him at Cornell in 1949. O. K. Bouwsma was there and met Wittgenstein for the first time, finding him "an attractive man with an easy and a friendly manner." But in discussion two days later he saw another side: "There is an intensity and an impatience about him which are enough, certainly, to frighten one, and there was [an occasion] when Norman was floundering, and going on talking . . . , when he was nearly violent."

Sometimes Wittgenstein's intense reaction moved beyond a display of intellectual ferocity to involve the violent shaking of a stick—or a poker. In 1937, on one of his retreats to Norway, Wittgenstein was puzzled by a change in his hitherto warm relations with a neighbor, Anna Rebni. A "tough, elderly Norwegian farmer," in Ray Monk's phrase, she became cold and distant toward him. Eventually he asked for an explanation and, so he recorded, could never have guessed her answer: "He had threatened her with his stick." But, as he went on to elucidate, it was his "habit, when I like someone very much and have a good relationship, that in a jolly mood, in the same way I might pat someone on the back, I make threatening gestures with a fist or my stick. It is a kind of cuddle." *(". . . mit der Faust oder dem Stock zu drohen. Es ist eine Art der Liebkosung.")*

As Wittgenstein's primary-school pupils would have testified, he was not slow to lash out at a head or an ear—sometimes making them bleed. A cuff to a pupil's head featured in his 1930s confessions. To return to H3, a contemporary witness of Wittgenstein's handling of pokers was Noel Annan, who went on to become provost of King's and to be given a seat in the House of Lords. Annan was a historian but, probably in that period, attended an MSC meeting with the guest speaker, J. L. Austin, the Oxford linguistic philosopher, whom he had known at General Eisenhower's Supreme Headquarters during the war.

> At some point Richard Braithwaite said something and I noticed that Wittgenstein reached down into the fireplace, picked up a poker and gripped it very tightly. He said "BRAITHWAITE YOU ARE WRONG" and the whole place was electrified. He didn't threaten Braithwaite, but I remember the incident because of the other happening about which Wittgenstein was famous. . . . In those days when you had coal fires there was always a poker around. So it was a perfectly natural thing to reach down and seize something on which you could express your anger.

Natural to Wittgenstein, perhaps.

Friedrich von Hayek saw Wittgenstein in action with the H3 poker at a meeting of the MSC to which he was taken by Braithwaite in the early 1940s:

> Suddenly Wittgenstein leapt to his feet, poker in hand, indignant in the highest degree, and he proceeded to demonstrate with the implement how simple and obvious "matter" really was. Seeing this rampant man in the middle of the room swinging a poker was certainly rather alarming, and one felt inclined to escape into a safe corner. Frankly, my impression at that time was that he had gone mad.

And there is one further piece of evidence of Wittgenstein's capacity for violence, from his own writing:

> When I am furious about something, I sometimes hit the ground or a tree with my stick, and the like. But I certainly don't think that the ground is to blame or that this hitting can help at all. I give vent to my anger. And that is what all rites are like. . . . The important thing is the similarity with an act of punishing, but nothing more than similarity is to be found.

For Wittgenstein, nothing more than similarity with the act of punishing. But what about for the recipient? In *Unended Quest* Popper is inclined to laugh off the brandishing of the poker. Eyewitnesses are understandably loath to accuse the great philosopher of a violent act. But the fact is that, in giving his example of a moral principle, Popper used the word "threaten:" "Not to threaten visiting lecturers with pokers." The exactness of the phrase "threaten visiting lecturers" is the one thing on which all the witnesses who heard the remark agree. And both his choice of a principle and his unthinking use of the word "threaten," even in a joke, point decisively to how Popper felt at the time and how personal the clash had become. The poker was real enough. So, it seems, for Popper, was the threat.

17

Trajectories of Success

He has the pride of Lucifer.
— RUSSELL ON WITTGENSTEIN

I think that success in life is largely a matter of luck. It has little correlation with merit, and in all fields of life there have been many people of great merit who did not succeed.

— POPPER

OF WHOM CAN POPPER have been thinking when he wrote those last words in the 1970s?

If there was one key difference between the lives of Wittgenstein and Popper, it was the trajectory of their careers. This provides another clue to the confrontation and whether Popper misrepresented it. Despite all the quirks of personality that made him so difficult a colleague, Wittgenstein always found sources of support at Cambridge, whereas Popper for many years found himself the odd man out of the academic world. It was his misfortune that the most creative part of his life was lived in Wittgenstein's shadow.

Wittgenstein's relationship with the academic establishment could well be described as one of love-hate: Cambridge was enthralled by him, but he barely tolerated the ancient university, even though he was to return, on and off, until his death in 1951.

Virtually from the hour of his unannounced arrival at Bertrand Russell's door in Trinity in 1911, he was recognized as exceptionally gifted and became all the rage in Cambridge social and intellectual society. In 1912, and over Russell's protests, he was invited to join the Apostles, described by the early socialist Beatrice Webb as being for those "who aim at exquisite relations within the close circle of the elite." Russell foresaw Wittgenstein's distaste. At that time the Apostles were dominated by the homoerotic Bloomsbury set—Maynard Keynes, Lytton Strachey, Rupert Brooke. Russell was accused of trying to keep the handsome Austrian genius to himself. Wittgenstein, however, as Russell predicted, was repelled by the precious and self-congratulatory tone of the Apostles' meetings. He dropped out, but most unusually—a mark of how much he was sought after—was reelected when he returned to Cambridge from Vienna in 1929. Keynes hosted the dinner in his honor; a fellow guest was Anthony Blunt.

How much Wittgenstein's homosexuality/sexual ambiguity was responsible for the Apostles' unusually generous attitude to him can only be a matter of speculation, just as how active a homosexual he was remains uncertain and a matter of controversy. That he had intense feelings toward a number of young men is clear. Like David Pinsent and Francis Skinner, they tended to be academically bright, immature, and disabled. That he was also tormented by those feelings, and by questions of what love meant, is also clear. The objects of his obsessions—the working man Roy Foureacre, for instance, whom Wittgenstein met at Guy's—sometimes appeared to be totally unconscious of his involvement with them. His supporters raged at William Warren Bartley III, who pictured him in a biography cottaging in the Vienna woods. As neither Wittgenstein's nor Popper's sexual orientation is directly relevant to our story, the matter can be left there.

In 1929, back in England, Wittgenstein submitted the slender, 20,000-word, *Tractatus Logico-Philosophicus* for his Ph.D. One of his examiners, G. E. Moore, is reported to have written on the form for the Ph.D. submission, "It is my personal opinion that Mr. Wittgenstein's thesis is a work of genius; but be that as it may, it is certainly well up to the standard required for the Cambridge degree of Doctor of Philosophy." A decade later, when Moore retired, as we have seen, even dons who disagreed with Wittgenstein's approach felt that it was impossible not to appoint him to the vacant chair.

This eminence came to him in spite of his never publishing any other substantial work of philosophy beside the *Tractatus*. In his lifetime, Wittgenstein published only two other works: his widely used German glossary for schoolchildren, and the script of a planned lecture to a meeting sponsored jointly by the philosopher's journal *Mind* and the Aristotelian Society. (He actually lectured on another subject, much to his audience's surprise.) The many volumes that now crowd the shelves of bookshops and libraries under his name were all compiled posthumously from his notes. In his book *Frege: The Philosophy of Mathematics*, Michael

Dummett comments sourly on the modern practice of assessing academic performance by indicators measuring the number of words published each year: "Wittgenstein . . . would plainly not have survived such a system."

Despite all this, and although his peers never questioned his brilliance, originality, and profundity, at the time of his death Wittgenstein remained barely known beyond philosophy. He was awarded no medal or title from the Crown; he was not summoned to meet international dignitaries; he gave no keynote addresses; there were no ceremonies or convocations in his honor—and it is unlikely that he would have agreed to any such thing.

Compare this to Popper's situation. Popper wrote and published prolifically. British newspapers carried fulsome tributes after his death. And it was noticeable that elsewhere in Europe the news was accorded even greater prominence—one leading Swiss paper, for example, devoted five pages to his life and works. Indeed, during his lifetime Popper was more revered abroad than at home, with prizes and honors across Europe and in America and Japan. By the time he died he had collected a score of honorary doctorates.

Wittgenstein's influence is on philosophers and artists; Popper's on the practical world of business, politics, and science. A winner of the Nobel Prize for Medicine, Sir Peter Medawar, described him as "incomparably the greatest philosopher of science that has ever been."

The billionaire Hungarian-born financier George Soros, a former student of Popper's, was so inspired by his teacher that he named his Open Society Foundation in his honor. Soros made his millions in the stock market investing on Popperian lines. Popper believed that the scientific theories that should be considered the most robust were those that offered themselves up to, and then survived, the most severe of tests. Soros applied this principle to Wall Street. Thus he made a fortune investing in a California mortgage insurance stock that had taken a hammering in a housing slump; the fact that it survived the recession, he thought, was strong evidence that the company was essentially sound. The Open Society Foundation was Popper's political theory put into practice, testing the transforming power of openness. The foundation provided money for books and scholarships, photocopiers and fax machines, debating societies and conferences, anything that facilitated and stimulated the exchange of ideas. So vital did it become in eastern Europe as a source of funds that it could certainly lay claim to having accelerated the collapse of communism. The German Social Democrat chancellor Helmut Schmidt wrote the foreword to a *Festschrift* for Popper: "Like no one before, he has illustrated with brilliant acuity the defects of the utopian state in his critiques of Plato, Hegel, and Marx—who, on the basis of strict and supposedly absolute premises, attempted to predetermine the course of political development." Popper was also a great hero to Schmidt's Christian Democrat successor, Helmut Kohl. The German president Richard von Weizsäcker went to Kenley to see Popper while on a state visit to Britain. The Czech president, playwright, and former dissident Vaclav Havel invited him to his palace in Prague. The Dalai Lama visited him, and the Japanese emperor invited him to the Imperial Palace. The Austrian chancellor Bruno Kreisky sent him "warmest congratulations" on his eightieth birthday. It is said that, when ninety, he was even sounded out to succeed the internationally ostracized Kurt Waldheim as president of Austria; if so, he laughed it off.

In Britain, too, tributes and honors were heaped upon him in later life—among them a knighthood and appointment as a Companion of Honour. He was said to be Margaret Thatcher's favorite philosopher: she wrote that Popper and von Hayek were her gurus. Though he had strong views on current political issues, and was willing to add his name to letters of protest, Popper always declined to participate in active politics. He began his adult life flirting with communism. Some thought he ended as a conservative, others said he remained essentially a social democrat in the European tradition. His appeal to the left lay in his concern for the worst-off in society and in his recognition of the need for government action to bring about social justice. His appeal to the right lay in his individualism, his rejection of heaven-on-earth visions, and in his distrust of policies designed to bring about rapid and large-scale transformation of any kind. He believed in change, but change brought about by piecemeal social engineering. His critique of communism and fascism attracted all democratic parties whatever their hue. And he supplied both the parliamentary left and the parliamentary right with a rationale for the free society.

And yet it had all been such an effort, such a hard and wearying journey. While he was making his way as a philosopher, attending conferences and giving lectures, he and his wife kept going by teaching in Viennese schools. It was only at the age of thirty-five, after turning his back on Austria in 1937 with the help and backing of Friedrich von Hayek, then at the London School of Economics (and, curiously, Wittgenstein's second cousin on his mother's side), that he took up his first full-time lecturership—in New Zealand, hardly the beating heart of philosophy.

In 1936, with the political portents becoming impossible for him to ignore, Popper had applied to the Academic Assistance Council in England for help in leaving Austria, complaining that he was the subject of anti-Semitism from his pupils and his fellow teachers.

The next stage was drawn out. The Council sent him a detailed form on which, among other information, he had to declare his earnings (two pounds a week) and tick a box to say whether he would be willing to be relocated to a tropical country in the British Empire ("Yes, if the climate was not too bad"). He had to prove to a well-intentioned British academic who helped run the AAC, Professor A. E. Duncan Jones, that, although he had not yet been forced to abandon his teaching post, he and other people of Jewish origin were now in real danger. Professor Duncan Jones even suggested, in internal correspondence, that Popper be encouraged to commit some "political indiscretion" to ensure his sacking, which would facilitate and expedite financial assistance from the AAC.

References were required and, given the gulf that Popper portrayed between him and the Vienna Circle, it is curious that when he took the decision to flee the city he called upon several members of the Circle, including Carnap, Kaufmann, and Kraft, to provide them.

It is also a measure of the man that at thirty-four, with one publication in German, there was a superstar quality to the referees he gave the AAC. It would be difficult to imagine a more celebrated list: there in Popper's handwriting are Albert Einstein and Niels Bohr alongside Bertrand Russell, G. E. Moore, and Rudolph Carnap (several of these Popper had impressed at conferences and lectures in the mid-1930s). Even so, Duncan Jones needed convincing that Popper was worthy of sponsorship. Various people he had approached on the Council's behalf, he wrote, informed him that Popper himself was not in the first rank of Viennese philosophers. But then Popper was still only a secondary-school teacher.

Another reference might have caused Popper even more heartache. The biologist Joseph Needham wrote to the Academic Assistance Council recommending Popper and suggesting that he would not be too much of a drain on their resources:

> In sum, there can be no doubt that given the chance to develop his work and publish it, he will certainly find a position some-where, since he is of the same type as Dr. Wittgenstein, who for some time past has been a Fellow of Trinity College here. The AAC could therefore be assured that their support would not be needed for more than a limited period.

Eventually all the pieces slotted into place, although to Popper the entire procedure must have felt dangerously slow. After securing the backing of the AAC he was offered a temporary lectureship at Cambridge. At one stage the formal offer of this "academic hospitality" appears to have been mislaid in the post; Popper dashed off a nervous letter to G. E. Moore begging him to send another.

By the time this came through, however, Popper had heard that his bid for a permanent lecturership at Canterbury University College, Christchurch, New Zealand—for which he had applied on 25 October 1936—had also been successful. He wrote again to the AAC, this time in German. He was a very happy man, even though New Zealand seemed a long way away—"It [New Zealand] is not quite the moon, but after the moon it is the furthest place in the world!"

It was not so far away that Wittgenstein was absent from his mind there. The index to *The Open Society and Its Enemies* contains fifteen references to Wittgenstein—all hostile—and in the notes there are pages of criticism of the *Tractatus*. In H3 he would have the first and only chance to repeat them to Wittgenstein's face.

IT WAS UNDOUBTEDLY a problem for Popper's reputation in the English-speaking world that his seminal first work, *Logik der Forschung*, though published in German in 1934, did not appear in English until a quarter of a century later. And, whatever its later fame, Popper had a long struggle to find a publisher for *The Open Society and Its Enemies*. Wittgenstein's *Tractatus* had its birth pangs, but mainly because of doubts about potential sales rather than about its quality. Popper's struggle was to convince the publishers of the ground-breaking nature of his endeavor. Without the efforts of his close friend and supporter Ernst Gombrich, working tirelessly on his behalf in London while the author grew increasingly frantic in New Zealand, the book might never have appeared.

In 1945, with von Hayek's help, Popper was offered a readership at the London School of Economics, and for twenty-three years from 1949 he held the chair of Logic and Scientific Method there. Arriving in England, he was lauded and fashionable: *The Open Society* had just been published, and invitations to speak flooded in. In retrospect, it looks like the peak of his professional life—which would continue for more than four decades—for soon he began to withdraw from conferences and professional gatherings, choosing increasingly to work in isolation at home. He became the "big man not present." The work would continue, edging into new areas, but the heroic period, tackling the fundamental questions, was over.

Popper's relationship with the philosophical establishment in his adopted country was always chilly; from early in his career he may have despaired of British audiences for their inability to appreciate his originality. This made him all the more assertive with them. Visiting Britain in 1936, Popper was present as A. J. Ayer's guest at a meeting of the Aristotelian Society where Russell was the speaker. He intervened, but the audience took his remarks as a joke, greeting them with laughter and clapping. Forty years later he wrote, "I wonder whether there was anybody there who suspected that not only did I seriously hold these views, but that, in due course, they would widely be regarded as commonplace." On this trip he also made his first foray to Cambridge and the Moral Science Club, where his paper dealt with induction and he narrowly missed an encounter with Wittgenstein, who was laid up in bed with what he described in a letter to G. E. Moore as "a bloody cold." For Popper, already preoccupied by Wittgenstein, that absence was a boon: ten years later, he himself would be a heavier gun.

The London School of Economics was, and is, justly renowned as one of Britain's preeminent institutions of higher education; a teaching post there is a recognition of professional excellence. But Popper never gained the Oxford or Cambridge appointment that he felt should have been his, seeing himself condemned to the role of perpetual outsider, deprived of full recognition of his talent. (John Watkins, however, challenged the view that Popper ever coveted an Oxbridge chair.) He received honorary doctorates from both Oxford and Cambridge, but that is hardly the same thing. In 1947 Stephen Toulmin turned down an offer from Popper to teach at the LSE because in those days Oxford and Cambridge were the only places to be. Popper no doubt understood that very well.

In the end, however, what placed Popper at odds with the British philosophical establishment was his fundamental standpoint: that the study of problems is worthwhile, the study of puzzles trivial. Bryan Magee, who has done as much as anyone to promote understanding of Popper's importance, says that this left Popper marginalized: "Because he believed this, and practiced it, always from outside the main thought-streams of the age, he was never in the fashion. And because he spent so much of his time attacking, and severely, the ideas of people he disagreed with, he was never popular." Popper himself once commented to Melitta Mew that Oxford, where the linguistic approach was dominant, had 150 philosophers and no philosophy. And John Watkins contrasted Popper's method with that prevailing in English universities: "He liked to go for big, clear, strong problems. About these problems he liked to have strong theses—maybe held tentatively at first and to be changed later. Now that's not your average senior lecturer in philosophy in the University of Birmingham. He takes some little concept, he says it's got this little bit of ambiguity about it . . ." And here Watkins rubbed together his thumb and forefinger to illustrate the pettiness of the attitude.

Of course, Popper had his disciples, centered in the LSE. To John Watkins, Popper was a great man, "with some rough edges . . . [but] up at a level most people are not at." Lord Dahrendorf felt Popper "was something very special." Ernest Gellner credits Popper with providing the clue to the most valuable things we possess, namely knowledge and freedom. But Pop-

perians hardly made up a school. In part this was because Popper
tackled one specific problem at a time, whereas Wittgenstein provided a method, a universal approach. Although Popper contributed to an impressive array of topics, it is rarely illuminating to
ask "What would Popper say here?" about those he ignored. But
the equivalent Wittgenstein challenge, so say the Wittgensteini-
ans, always throws up a response.

There was also another reason for Popper's relatively poor reception in England and much higher profile elsewhere. It lay in
his intellectual common sense—an admirable quality, but not a
fascinating one. Ralf Dahrendorf, much of whose academic career was in Britain, saw the reactions to Popper from both sides:

> Popper was very happy in England because he felt safe. It was a
> country in which a man who was immune to the great passions of
> the century—notably communism and fascism—could feel that
> he wasn't challenged. But, precisely because England was such a
> country, Popper was too normal to be interesting. Now the Conti-
> nent has the opposite story. The great passions threatened every
> single country. And there Popper stood, a tower of reason in the
> midst of turmoil. And that, over time, commanded enormous re-
> spect. More than that, it was regarded as the great answer to the
> destructive and disastrous consequences of the passionate policies
> from 1917 to Stalin's death, and that includes the whole of the
> Nazi period.

Did Popper ever regret taking up the full-time post in New
Zealand, far from Nazi Austria and far from the war, rather than
gambling on an insecure offer from the Academic Assistance
Council of "academic hospitality in the name of the Moral Sci-

ences Faculty of Cambridge University?" The University of Canterbury certainly benefited. Its official history reports, "Popper's impact on academic life was greater than that of any other person before or since." The New Zealand historians wrote that he acted as a kind of "intellectual champagne after the dry depression years."

Yet if he had gone to England in 1937 he would not have spent some of his most productive years outside the mainstream of philosophy. He would have had the chance to establish himself academically—and to work and debate alongside Wittgenstein. Friedrich Waismann, his fellow Viennese, to whom Popper claimed to have handed the Cambridge opportunity, moved from Cambridge to Oxford. How Popper would have relished that! Had he gone to Cambridge, he remarked to Michael Nedo, Wittgenstein and his school would have been eclipsed.

When the New Zealand offer, from the Universities Bureau of the British Empire, arrived in Vienna, he wrote to G. E. Moore that the letter had been "rather unexpected." But "I am very happy to obtain this position. For although I would certainly prefer to have the opportunity of lecturing in Cambridge I am satisfied not to be a burden to the Academic Assistance Council any longer—and I hope, once to have the possibility of coming back to England."

In fact the choice in favor of what Popper called, significantly, a "normal job" can be understood only against a background of personal, financial, and political insecurity. Predicting war, he took the decision to leave Austria over a year before Hitler invaded. And we should remember that the year he completed his Ph.D., 1928, and then qualified to teach, was the last year of Austrian postwar recovery. The American stock market collapsed in 1929 and capital flooded out of Europe. In 1930 German unemployment reached over 5 million (it would be 6 million two years later). With the state of the economy, with political tensions, the rise of right-wing parties, and vicious anti-Semitism, in that year the would-be teacher of Jewish origin started a career and a marriage on the most precarious of foundations. The world had shifted for the boy born into the certainties of Jewish achievement.

Radiator as work of art. Designed in 1928 by Ludwig for Margarete's Kundmanngasse house. His requirements were so exacting that no Austrian firm could do the casting.

The cabinetmaker. As a young man, Popper wanted, like Wittgenstein, to be a manual worker and became apprentice to a cabinetmaker. In 1924, with this hanging cupboard, he completed his apprenticeship. He wrote that he learned more about the theory of knowledge from his apprentice-master "than from any other of my teachers."

18

The Problem with Puzzles

The lord whose oracle is in Delphi neither speaks nor conceals but gives a sign.

—HERACLITUS

The later Wittgenstein used to speak of "puzzles," caused by the philosophical misuse of language. I can only say that if I had no serious philosophical problems and no hope of solving them, I should have no excuse for being a philosopher: to my mind, there would be no apology for philosophy.

—POPPER

AS WE HAVE SEEN, there were many factors at work in H3 to make the encounter between Wittgenstein and Popper so fierce. But the debate might still have been memorable without them, for at stake was the most fundamental issue in philosophy: its very purpose. Linked with this was the custodianship of the analytic revolution, which had been led by Bertrand Russell. It was over the meaning and direction of this philosophical insurrection that Popper and Wittgenstein would, metaphorically at least, cross pokers.

At issue was the importance of language. Russell had initiated the rigorous use of the techniques of logic to analyze philosophical problems. Until Russell, and from the time of Descartes in the seventeenth century, the central branch of philosophy had been epistemology—the study of what we can know. Descartes had searched inside himself for secure knowledge. His method had been to doubt everything until he reached the bedrock of certainty. When his spade could dig no further, he coined philosophy's most oft-quoted phrase, "Cogito ergo sum"—"I think, therefore I am." The British empiricists Locke, Berkeley, and Hume were among those who followed in this epistemological tradition. But, after Russell, epistemology was displaced by the philosophy of language and the premise that our words are the lenses through which we access our thoughts and the world. We cannot see the world without them.

Russell's analytic approach had its origins in numbers; mathematics was his first love. In his autobiography he recalled his miserable adolescence and a footpath down which he would wander on England's south coast. "I used to go there alone to watch the sunset and contemplate suicide. I did not, however, commit suicide, because I wished to know more about mathematics."

In 1903 he published *The Principles of Mathematics*, and in 1910–13, with a coauthor, A. N. Whitehead, the monumental three-volume *Principia Mathematica*. This bid to place mathematics on a secure logical footing required hundreds of pages of numbers, symbols, and equations: the result was so unmarketable that the authors had to contribute toward the publishing cost. Russell was later to claim that he knew of only six people who had read it from beginning to end: three who perished in the Holocaust and three from Texas. However, it indirectly paid for itself by giving authority to his more populist writings.

Russell was ultimately to regard the task attempted by *Principia Mathematica* as more Sisyphean than Herculean—or "unmitigated rubbish" in his words—and the real significance for philosophy came when he transferred the techniques he had employed in this work to the study of language and then to the perennial problems of metaphysics: the nature of existence, knowledge, and truth. The most famous of his theories concerns the baldness or otherwise of the French monarch—but the debate about this nonexistent hairless cranium can be understood only in the context of the philosophical fixations of the time.

A great puzzle for philosophers was the relationship between language and the world. How was it that a series of letters, say *p-i-p-e*, when placed in the appropriate order, acquired a meaning? The creed in the early twentieth century—part of the branch of philosophy known as logical atomism—was that all words stand for objects and that that is how a word derives its meaning. The word "pipe" stands for the object pipe; the word means the object.

But this view of the link between language and the world raises a number of perplexing issues. For example, what object does a fairy-tale creation such as a golden mountain signify? Clearly we can construct perfectly ordinary and comprehensible sentences involving a golden mountain. We can even make sense of a statement such as "The golden mountain does not exist." And yet this statement is odd, for we are apparently referring to an object, a golden mountain, only to go on to deny that there is an object that can be referred to. If I ask, "What is it that does not exist?" the answer "The golden mountain" appears to confer some sort of reality upon the nonexistent mountain.

There was a related puzzle. If the meaning of the name "Sir Walter Scott" is the object or thing denoted by the name—in other words, the person Sir Walter Scott—then presumably this is also true of a description of Sir Walter Scott such as "the author of *Waverley*." The description "the author of *Waverley*" also denotes Walter Scott; consequently it must mean the same as the name. Yet this account of descriptions lands us in further difficulties. For when King George IV wanted to know whether "Scott was the author of *Waverley*," he presumably did not want to know whether "Scott was Scott." As Russell so neatly put it, George IV was not expressing an interest in the law of identity.

Finally, back to our bald Frenchman. Even though France is a republic, we have no difficulty in understanding the sentence "The King of France is bald." It is perfectly coherent. One might utter this sentence at a party, and a person not versed in the French constitution might believe it to be true. In this sense it is not at all like the nonsense sentences "King is a bald France" and "France bald is King a."

But how is it that we can comprehend the sentence "The King of France is bald" when there is no King of France? If "the King of France" referred to a person, then this person would have to be either bald or not bald, just as the poker in H3 must have been either red-hot or not. Yet out in the real world there is not a single hairless person who is the King of France, nor is there a single hirsute person who is the King of France. Russell wrote that, "Hegelians, who love a synthesis, will probably conclude that he wears a wig."

Writing at the turn of the nineteenth century, an Austrian logician, Alexius Meinong, had proposed a response to these problems. For Meinong, the fact that we can refer to a golden mountain means that there is a way in which a golden mountain is out there—not physically out there, of course, but logically out there. The same is true of unicorns, Easter bunnies, tooth fairies, ghosts, goblins, and the Loch Ness monster. That is how it is possible for us to make sense of the claims "Father Christmas does not exist" and "The Loch Ness monster is nothing but a big trout." In the world of logic there is a Loch Ness monster. Its existence in the world of logic allows us to deny its existence in the world of reality.

Now Russell was a very methodical and meticulous man. (During 1946 he was occupying C. D. Broad's rooms in Trinity Great Court. Broad later noted, "I am glad to record that, however destructive he may have been as a thinker, he appeared on my return to have been a model tenant.") The picture of the world conjured up by Meinong seemed to Russell intolerably cluttered and disorderly. "Logic," he thought, "must no more admit a unicorn than zoology can." And it was to spring-clean this metaphysical mess that he invented his ingenious Theory of Descriptions.

We are confused by our language, Russell believed. We think that descriptions such as "the golden mountain" or "the author of *Waverly*" or "the King of France" behave like names. In a crowd awaiting a royal procession, we might equally exclaim "Finally, here's the King of France" as "At last, it's Louis XVIII." And so we think that these descriptions, just like names, must denote an object in order to have meaning.

But in fact they do not function as names at all. Although the statement "The King of France is bald" seems straightforward, it actually masks a complex logical triplet. It is a cheese and tomato omelette disguised as an egg. And its three ingredients are as follows:

1. There is a King of France.
2. There is only one King of France.
3. Whatever is King of France is bald.

Once this logic is exposed, we can see how the statement "The King of France is bald" makes sense but is false: it is because the first premise, that there is a King of France, is untrue. "The golden mountain does not exist" and "Scott is the author of *Waverley*" can be treated in much the same way. "Scott is the author of *Waverley*" is translatable as "There is an X, such that X wrote *Waverley*, such that for all Y, if Y wrote *Waverley*, Y is identical with X, and such that X is identical with Scott."

Russell invented a logical notation to deal with these cases — one that is still used today. "The King of France" is bald is expressed as

$$(\exists x)[Fx \,\&\, (y)(Fy \longrightarrow y=x) \,\&\, Gx]$$

This deconstruction of the sentence has come to be seen as a paradigm of the analytic method. And thereafter, when asked what was his most significant contribution to philosophy, Russell would unhesitatingly answer, "The Theory of Descriptions."

The bald French monarch first reared his gleaming head in an article Russell published in 1905. Four decades later, in room H3 on 25 October 1946, Russell, the parent to this novel methodology, was sitting flanked by two philosophic offspring, Popper and

Wittgenstein. Like children in so many families, they were at war. Wittgenstein had come to see the linguistic scrutiny of concepts as of value in itself—the only pill we need to swallow to ease our philosophical migraines. For Popper it was no more than an extremely useful device in the examination of what mattered—real problems.

SINCE RETURNING TO CAMBRIDGE IN 1929, Wittgenstein had abandoned most of the ideas contained in the *Tractatus* and had developed a radically new approach. While very few people in the history of philosophy can boast the creation of one school of thought, Wittgenstein can lay claim to the foundation of two. Russell labeled the two approaches Wittgenstein I and Wittgenstein II.

The Wittgenstein of the *Tractatus* had been working in the same intellectual universe—logical atomism—in which Russell had carried out his early and most original work, where the world is constructed on simple, unchanging (and indefinable) objects. The text of the *Tractatus* is sandwiched between its well-known opening and closing statements: "The world is all that is the case" and "Whereof one cannot speak, thereof one must remain silent." It is written in intriguingly numbered paragraphs, 1 to 7, with decimal numbers reflecting the relative significance of their subclauses, 1.0 being more significant than 1.1, which in turn is more significant than 1.11, and 1.111. For example:

4	A thought is a proposition with a sense.
4.001	The totality of propositions is language.
4.01	A proposition is a picture of reality . . .

> 4.1   Propositions represent the existence and non-existence of states of affairs.
>
> 4.1212   What *can* be shown, *cannot* be said.

The monograph is remarkable for its combination of luminosity and oracular brevity, its impregnable confidence teetering on the edge of dogmatism (Popper certainly thought it dogmatic), and its unwillingness to demean itself by supporting its numbered propositions in any conventional way. The individual sentences have a pure and simple beauty—indeed, *The Times*'s obituary of Wittgenstein described the *Tractatus* as "a logical poem."

Central to the project is the connection between language, thought, and the world. In particular, Wittgenstein offers a picture theory of meaning: facts and propositions, such as "The fireplace is in the center of the room," somehow present a picture of the way the world is. This is an idea which had dawned on him after reading in a newspaper about a court case in France in which models of automobiles and pedestrians had been used to demonstrate what happened in a real accident. Propositions stand in a similar relation to the world as the display of toy cars and dolls stood in relation to the accident.

But in Wittgenstein II the metaphor of language as a picture is replaced by the metaphor of language as a tool. If we want to know the meaning of a term, we should not ask what it stands for: we should instead examine how it is actually used. If we do so, we will soon recognize that there is no underlying single structure. Some words, which at first glance look as if they perform similar functions, actually operate to distinct sets of rules. It is like peering into the cabin of a locomotive. We see handles that all look more or less alike.

But one is the handle of a crank which can be moved continuously (it regulates the opening of a valve); another is the handle of a switch, which has only two effective positions, it is either off or on; a third is the handle of a brake-lever, the harder one pulls on it, the harder it brakes; a fourth, the handle of the pump: it has an effect only so long as it is moved to and fro.

Moreover, if we examine how language is actually used, we will notice something else: that most terms have not just one use but a multiplicity of uses, and that these various applications do not necessarily have a single component in common. Wittgenstein gave the example of the term "game." There are all sorts of games—patience, chess, badminton, Australian-rules football, kids playing catch. There are competitive games, cooperative games, team games, individual games, games of skill, games of luck, games with balls, and games with cards. Question: what is it that unites all the games? Answer: nothing. There is no essence of "game."

Wittgenstein called such terms "family resemblance" concepts. They are like a family, some members of which might have the distinctive family craggy neck, or piercing blue eyes, or premature white hair, or an unusually large pair of ears, but where there is not a single characteristic possessed by all. What makes "games" games is an overlapping series of similarities and resemblances. It is this very crisscrossing which gives concepts their stability. In this they resemble a thread, "where the strength of the thread does not reside in the fact that some one fibre runs through its whole length, but in the overlapping of many fibres."

Russell and the early Wittgenstein believed that everyday language obscures its underlying logical structure. "The King of France is bald" is a proposition whose logical structure is not immediately apparent on the surface. Language is a covering, like clothing to the body. A baggy jumper may disguise the shape inside. Wittgenstein II did not take this view: he believed that language is in perfect working order—it hides nothing.

The later Wittgenstein held that, instead of language being somehow chained to the world of objects, grammar is autonomous—it runs free. We, not the world, are the masters. We can do with language what we wish. We chose the rules and we determine what it means to follow the rules. Over the next few decades these ideas were to infiltrate the study of law, sociology, and English around the world.

Since language is governed by rules, it is also essentially public; it is embedded in our practice, in our "forms of life." Rules have to be interpreted; there has to be consensus on what is permissible and what is not. Thus the idea of a private language—a language that only one person can understand—is incoherent. And if this is correct then Descartes, by looking inside himself for incontrovertible knowledge, had sought the holy grail of certainty from the wrong direction. If *"Cogito ergo sum"* is to have any meaning, there has to be prior acceptance of what is to count as thinking, and how the concept of "thought" is to be used—for that is the only way in which language can function. It is therefore quite impossible that the *Cogito* can be the starting point of what we can know. With this insight, Wittgenstein overturned several hundred years of philosophy and emancipated his followers from the slavery of the search for rock-bottom certainty.

What then, for Wittgenstein, was the aim of philosophy? Quite simply, to disentangle ourselves from our self-enveloped confusion—"to show the fly the way out of the fly bottle." When we engage in philosophy, we puzzle about things that ordinarily do not concern us. What, for example, is the nature of time—if it is five o'clock in Cambridge, is it also five o'clock on the sun? Can something be both red and green all over? Can I know I am in pain? Can I have the same pain as you? What is it to speak to one-self? (That was the question broached at Wittgenstein's afternoon seminar on 25 October 1946.)

In seeking answers to these questions, Wittgenstein II believed, philosophers make foolish errors. They look for an explanation, a universal answer, a theory to cover all cases, a generalization to fit all types; they stare at objects and feel they can somehow pene-trate phenomena and reach an immaterial core.

Such philosophizing may sound a little like incipient de-rangement, and in fact Wittgenstein II conceived philosophy as a sort of linguistic therapy, a parallel to the approach of his sister's friend Sigmund Freud. "The philosopher's treatment of a ques-tion is like the treatment of an illness." Indeed, the 1946 secretary of the Moral Science Club, Wasfi Hijab, says that until he met Wittgenstein he was "intellectually sick," suffering from such con-fusions. Wittgenstein "cured" him.

What we must do, thought Wittgenstein II, is battle against the bewitchment of our language. We should constantly remind our-selves about everyday language—language in the home. Our baf-flement arises when language is used in unfamiliar ways, "when language goes on holiday." Can something be red and green all over? No, but that is not a deep metaphysical truth—it is a rule of our grammar. Perhaps in a far-flung corner of the world, in a dis-tant part of a remote jungle, there is an undiscovered tribe in which descriptions of shrubs or berries or cooking pots as "red and green all over" are commonplace.

Philosophical questions, then, are puzzles rather than problems. In unraveling them, we are not uncovering the hidden logic unearthed by Russell and Wittgenstein I, but merely reminding ourselves of what already exists, how language is actually employed. Can I "know" I am in pain? Well, in ordinary usage this is not a question that can be raised. Expressions of knowledge — "I know that Vienna is the capital of Austria," for example — are predicated on the possibility of doubt. But my pain is, to me, beyond doubt. What time is it on the sun? We cannot say — not because we do not know the answer, but rather because the concept of time on the sun has not been allocated a place in our language; there are no rules to govern its application.

Does all this mean that philosophy is useless except to those intent on earning their living by it — those liable to fall into the mire of self-deluded profundity? As Gilbert Ryle put it, what has the fly lost who never found himself in the fly bottle? The answer of Wittgenstein II was that his method combats the philosopher in us all. We are almost bound to topple into fly bottles — it comes with the language. Although only a few of us are philosophers lecturing at the podium, all of us are philosophers at the kitchen table or in the Dog and Duck.

THE *Tractatus* is still widely read, and some of its logical innovations, such as the deployment of "truth tables" to set out the conditions under which a sentence is true or false, are in use to this day. Nevertheless, Wittgenstein's current reputation and influence largely rest on his later work.

However, at least one thing unites Wittgenstein I and Wittgenstein II — a preoccupation with language. Wittgenstein I believed that our ordinary daily language is slapdash and that attending to the hidden structure of language will enable us to solve puzzles. In the preface to the *Tractatus*, Wittgenstein writes that so-called philosophical problems arise only because we misunderstand "the logic of our language." Wittgenstein II believed that attending to the surface of language can solve puzzles, and that our troubles arise when we try to burrow beneath this surface.

Linked to this lifelong fixation with language was the underlying goal of separating sense from nonsense. In Wittgenstein I this project is undertaken in a most rigorous fashion; in Wittgenstein II the highlighting of a proposition such as "X is red and green all over" serves much the same purpose. This looks like a sentence that has a meaning and can be understood, but in fact it differs subtly from ordinary basic propositions. It is akin to a pump in a locomotive cabin which one assumes must perform a function until one notices that it is disconnected from all other pieces of equipment. One of the aims of philosophy, thought Wittgenstein, is to turn latent nonsense into patent nonsense.

POPPER AVERRED THAT when he arrived in Cambridge on 25 October 1946, he was limbering up to confront the Wittgenstein I of the *Tractatus*, a book that he had picked over in every detail. (He must have been quick off the mark, since he records in *Unended Quest* that he read it "some years" before writing his Ph.D., which he began in 1925; the *Tractatus* was published in its German form in 1921.) However, wielding a poker before him was Wittgenstein II. There were good reasons for Popper not to know this. Until the end of 1945 he had been in New Zealand, whereas Wittgenstein's unpublished writings had only circulated samizdat-style among his disciples. In Cambridge their arguments and teasing aphorisms—"If a lion could talk, we couldn't understand him"—which seemed at once impenetrable and profound had become a dominant influence. Yet this had not yet spread to London, let alone nearly to the other side of the moon. Stephen Toulmin charges Popper with being engaged in "old forgotten far-off things and battles long ago."

No matter: the aspect of Wittgenstein I with which Popper was determined to take issue was also central to Wittgenstein II. What Popper opposed was the emphasis placed on language. A sharp footnote in *The Open Society* in which he dismisses Wittgenstein's *Tractatus* doctrine that the true task of philosophy is not to propound sentences but to clarify them could equally well be directed at Wittgenstein II.

That Popper was unaware of the extent of the revolution in Wittgenstein's thinking is confirmed by his former New Zealand student and eyewitness of that night, Peter Munz, and is further corroborated by a lecture that Popper delivered in 1952, a year after Wittgenstein's death. The title of that lecture was "The Nature of Philosophical Problems and Their Roots in Science," and in the published version Popper added a footnote:

> Wittgenstein still upheld the doctrine of the non-existence of philosophical problems in the form here described when I saw him last (in 1946, when he presided over a stormy meeting of the Moral Sciences Club in Cambridge, on the occasion of my reading a paper on "Are There Philosophical Problems?"). Since I had never seen any of his unpublished manuscripts which were privately circulated by some of his pupils I had been wondering whether he had modified what I here call his "doctrine;" but on this, the most fundamental and influential part of his teaching, I found his views unchanged.

(Note Popper's phrase "when I saw him last," implying a string of previous meetings, even that he and Wittgenstein went back a long way together, perhaps often sharing a *Stammtisch* in Viennese days—something to impress the reader. But, as we know, the occasion in H3 was the only time they met.)

Until his death, in what borders on obsessive behavior, Popper could not resist taking potshots at Wittgenstein. As far as he was concerned, the "existence of urgent and serious philosophical problems and the need to discuss them critically is the only apology for what may be called professional or academic philosophy." An early jab in *Unended Quest* comes when Popper is remembering his childhood: he begins that section by remarking, "I have long believed that there are genuine philosophical problems which are not mere puzzles arising out of the misuse of language. Some of those problems are childishly obvious." And there were more *ad hominem* attacks: "Wittgenstein . . . did not show the fly the way out of the fly bottle. Rather, I see in the fly unable to escape from the bottle, a striking self-portrait of Wittgenstein. (Wittgenstein was a Wittgensteinian case—just as Freud was a Freudian case.)" He cocks another snook in *Unended Quest* when talking about two Viennese authors in his father's library, Fritz Mauthner and Otto Weininger, "both of whom seem to have had some influence on Wittgenstein." The footnote to this then quotes Weininger: "All blockheads, from Bacon to Fritz Mauthner, have been critics of language."

In a BBC radio interview in May 1970, Popper was scathing about Wittgenstein's posthumously published work:

If you force me at gunpoint to say what it is I disagree with in Wittgenstein's *Philosophical Investigations*, I should have to say,

"Oh—nothing . . ." Indeed I only disagree with the enterprise. I mean, I do not disagree with anything which he says, because there is nothing with which one can disagree. But I confess I am bored by it—bored to tears.

In fact his response plainly went beyond boredom. Joseph Agassi commented, "Wittgenstein was the philosopher's only bête noire: there could be no greater expression of loyalty to him than to lunge at Wittgenstein." Popper compared the interest in language to the practice of cleaning spectacles. Language philosophers might think this is worthwhile in itself. Serious philosophers realize that the only point of the cleaning is to enable the wearer to see the world more clearly.

POPPER BELIEVED THAT Russell stood shoulder to shoulder with him in his criticism of Wittgenstein, and he believed correctly. The collapse in the personal relationship between Russell and Wittgenstein, already chronicled, was exacerbated by the hostility with which they now regarded each other's theories. Wittgenstein I may not have agreed with all the fruits of Russell's early logical and technical work, but he was fully engaged with it. Indeed the *Tractatus* was a project conceived partly to correct what Wittgenstein saw as Russell's errors, and Russell's presence can be felt in almost every sentence. The preface to this slim work pays generous tribute to "his friend Mr. Russell," and in the body of the text Russell is referred to twenty-eight times. In contrast, by the time the *Philosophical Investigations*—the book most closely associated with Wittgenstein II—came to be written, from the late 1930s on, the author appears to have performed a remarkable disappearing act on Russell. Wittgenstein mentions his philosophical mentor only twice, both times critically.

For his part, Russell believed that the new ideas being promoted by Wittgenstein were dragging Cambridge philosophy down a cul-de-sac of tedium and triviality. He later wrote that he found Wittgenstein II "completely unintelligible. Its positive doctrines seem to me trivial and its negative doctrines unfounded." As for the *Philosophical Investigations*, "I do not understand why a whole school finds important wisdom in its pages."

Russell had pioneered the analysis of concepts, and, like Popper, thought this could often clarify issues and clear away the fog which surrounded them. But, also like Popper, he believed precision was not the be-all and end-all. Popper pointed out that scientists managed to accomplish great things despite working with a degree of linguistic ambiguity. Russell averred that problems would not disappear even if each word were carefully defined. He told the following anecdote as an illustration. He was cycling to Winchester and stopped to ask a shopkeeper the shortest way. The shopkeeper called to a man in the back of the premises.

"Gentleman wants to know the shortest way to Winchester."

"Winchester?" an unseen voice replied.

"Aye."

"Way to Winchester?"

"Aye."

"Shortest way?"

"Aye."

"Dunno."

In his book *My Philosophical Development*, Russell dismissed Wittgenstein's later view that ordinary language was in great shape, and that our philosophical worries were merely puzzles, linguistic cramps: "We are now told that it is not the world that we are to try to understand but only sentences, and it is assumed that all sentences can count as true except those uttered by philosophers." Elsewhere he accused Wittgenstein of debasing himself before common sense. What passes for common sense, Russell thought, was often in reality just prejudice and the tyranny of custom. And if Wittgenstein was right, he argued, then philosophy was "at best, a slight help to lexicographers, and at worst, an ideal tea-table amusement." As Russell and Popper drank tea on the afternoon of 25 October 1946, just four hours before the MSC meeting, they would have agreed that there was much more to philosophy than that.

There was, for example, the real world of international affairs. To comprehend fully the ferocity of the debate in H3, we have to consider its political subtext. Remember the date: 1946. The threat of fascism had just receded. The Cold War had just begun. Was politics something with which philosophers should engage? For Popper and Russell the answer was unequivocally yes—though, unlike Russell, Popper would not have been found on marches and at sit-ins: he was a wielder of the pen not the sword. Indeed, his experience of seeing demonstrators shot in Vienna convinced him that victory was best won with the pen.

It is arguable that Popper was Marxism's most effective critic, shattering its scientific pretensions. According to Popper, valid science offers itself up to scrutiny and makes predictions that can be tested. The bolder the predictions the better. Pseudoscience—in which category Popper lumped both neo-Marxism and Freudian psychoanalysis—either declines to submit itself to any test (by failing to make clear predictions through which it could be falsified, as the theory of relativity offered the test of observation which was made by Sir Arthur Eddington) or makes predictions but then finds a way of explaining away apparently conflicting evidence. The revolution did not take place in the country with the most developed proletariat. "Ah well, but that's because . . ." Capitalism has not led to a greater concentration of wealth into fewer and fewer hands. "Ah well, but that's because . . ." The neo-Marxists are full of "Ah well." (But not Marx himself, whom Popper held in high regard and who did make predictions, though Popper considered them falsified.)

For his "war effort," Popper transplanted some of these ideas to *The Open Society and Its Enemies.* The book tracks the roots of fascism, pointing the finger of blame particularly at Plato and Hegel. But its critique of fascism is equally applicable to other forms of totalitarianism, and it is this which gives the book its immutable quality and its relevance to contemporary closed societies, be these religious-fundamentalist, extreme nationalist, or ethnic chauvinist. Although his target was the philosophy of totalitarianism, Popper did not much mind that many believed that *The Open Society* was written as a Cold War polemic directed mainly at Marxism.

In *The Open Society* he vanquished the notions that progress is inevitable and that history is governed by inexorable and discoverable laws. There is no plot to history, he insisted. "History cannot progress. Only we human individuals can do it." And, though nothing is guaranteed, the most effective fertilizer for social and economic advance is "openness," which is poison to totalitarianism. In the year 2000, a Chinese academic, Liu Junning, was evicted from the Chinese Academy of Social Sciences after delivering a lecture on *The Open Society*.

Popper's idea that progress comes through trial and error was one of the truly great ideas of the twentieth century and, like many truly great ideas, it had the mark of utter simplicity. Error was always possible; a "truth" was never certain. Einstein had demonstrated that with his bold hypotheses, overturning the apparently inviolable theories of Newton. And Einstein believed that his theory, in turn, was flawed and would one day be replaced. We should embrace falsification, thought Popper, for when a theory was shown to be defective new problems were thrown up, and this was how science evolved. (The price to be paid was insecurity over whether we have finally reached the truth.) Just as the possibility of falsification is what distinguishes true science from pseudo-science, so the need to test, probe, and scrutinize is what makes the open society essential if political advances are to be made. Popper's insight was to recognize that democracy should not be viewed merely as a luxury, something a country can afford only once it has reached a certain stage of development. Rather, democracy itself is a prerequisite to progress. He believed democracy entails a rational attitude that can be summed up in the maxim he frequently cited: "I may be wrong and you may be right, and by an effort, we may get nearer to the truth."

However, being able to choose a ruler is not a sufficient condition of democracy. Indeed, when Popper looked at Plato's question "Who shall rule?" he condemned it as dangerous. Our concern should not be with legitimacy. After all, Hitler came to power legitimately: the Enabling Act that empowered him to rule by decree was passed by a parliamentary majority.

In *[The Open Society]* I proposed replacing the Platonic question "Who shall rule?" with a radically different one: "How can we draw up the constitution in such a way that we are able to get rid of the government without bloodshed?" This question places the stress not on the mode of *electing* a government but upon the possibility of *removing* it.

For Popper, how we should be governed and how society should be structured were real issues, no less suitable as the focus of the philosopher's gaze than induction or the concept of infinity. Indeed, in obvious ways, the subject matter was even more pressing. And underlying Popper's detestation of Wittgenstein was disdain for the latter's apparent indifference to the burning questions in the real world, at least as subjects to which the philosopher can make a valid and special contribution.

Russell, a great admirer of *The Open Society*, was even more of a political animal than Popper. He too thought that philosophers should address themselves to contemporary controversies beyond the ivory tower. In 1946 his growing concern was the threat of a nuclear apocalypse. A year later he would deliver a series of lectures in Holland and Belgium, advocating a radical solution — world government, "possessed of a monopoly of all the more powerful weapons of war."

At this time, his third wife, Patricia (known as Peter), was backing a campaign to improve conditions for those living in the British zone of occupied Germany. A few weeks after the Popper meeting, on 18 November, she was a signatory to a letter to *The Times* after a government announcement that British citizens were to receive extra rations for Christmas. The letter complained that this came in a period of food shortages in the British zone. "We suggest that unless the Government adopts a radical change in its food polices, we are endangering not only the immediate stability of Europe but also the chances of a true peace."

Although the war was over, the future of Europe looked bleak. Industry lay in ruins, basic necessities were in short supply, Communist parties were flourishing in some Western democracies, the Soviets were strengthening their grip in eastern Europe and developing the bomb. These developments presented the West with immediate threats to its democratic future. Meanwhile, Popper and Russell frustratedly watched Wittgenstein persuade a generation of new philosophers that philosophy was solely, as they saw it, trifling with language. It was essential for the future of philosophy that this deception should be exposed.

19

The Puzzle over Problems

People say again and again that philosophy doesn't really progress, that we are still occupied with the same philosophical problems as were the Greeks. But the people who say that don't understand why this has to be so. It is because our language has remained the same and keeps seducing us into asking the same questions.

—WITTGENSTEIN

FOR POPPER—a philosopher in the grand tradition—the real problems with which his fellow philosophers should engage ranged from the structure of society to the nature of science, from the relationship between mind and body to the meaning of infinity, probability, and causation. Several of these topics were on stage in the H3 drama.

When Wittgenstein demanded examples of problems, it was inevitable that "induction" (Will the sun rise tomorrow?) would be among the first that Popper would cite. Popper's deployment of the problem of induction to attack the verification principle had been a cause of his strained relations with the Vienna Circle; it was also the topic on his only previous visit to the bear pit of the Moral Science Club, and it remained an obsession. He believed he had solved the conundrum, and it is reported that toward the end of his life any attempt to resurrect it could incite in him an exasperated rage, as though an attempt was being made to glue back together an idol he had shattered.

As already described, having rejected the Circle's principle of verification as flawed, because grounded on inductive reasoning, Popper used his alternative falsifiability thesis to distinguish not sense from nonsense, but the scientific from the nonscientific. This falsification principle, however, has itself taken something of a bruising from critics. It has been argued—by Popper's disciple-turned-enemy Imre Lakatos, for example—that some theories ought to survive falsification, indeed that some great theories have survived early falsification. There are occasions when it is the experiment you want to reject or explain away, rather than the hypothesis. Thus, when scientists test Galileo's theory of gravitational pull by dropping steel balls of different masses down a mine shaft, an apparent refutation of the theory is taken as evidence of the presence of an interfering factor such as iron ore. The theory is considered sufficiently robust not to be dismissed by one possibly anomalous result.

Moreover, claimed Lakatos, a hypothesis should not be judged simply on the basis of the number and boldness of its predictions. For what is particularly of interest are those unique predictions not made by other theories—since otherwise a test could simultaneously corroborate several theories. If you are stationary on earth and throw a stone, Einsteinian and Newtonian physics make approximately the same predictions as to where the stone will land, whereas they will make very different predictions if you throw it from a spaceship. But if this claim is right, science takes on a sort of subjective, sociological component: a theory is to be judged not just against the world, but against other conjectures swirling around at the same time.

Popper believed his theory could withstand such sniping. However, by far the most significant criticism of his work is that, despite his grand claim, he failed to solve Hume's problem of induction. Popper's critics insist he did not satisfactorily answer why, in Imre Lakatos's example, one should not jump off the top of the Eiffel Tower. It is true that the theory that gravitational pull will quickly bring you down with a splat has been tested by innumerable accidents and suicides. It is also true, as Popper pointed out, that one could not logically deduce that it would do so the next time someone jumped. Nevertheless, unless one believes that the past is at least some guide to the future, there is no reason not to take that leap.

Whether or not Popper satisfactorily addressed these objections, he never believed they could be dissolved in the analysis of language. Although he had already sketched out his approach to induction in *Logik der Forschung*, in the English-speaking world in 1946 it was barely known. At the foot of the letter he wrote to Russell two days after the Moral Science Club meeting, he offers to explain his solution to the 200-year-old problem of induction. He would not need much of his hero's time. It would take, he says, just twenty minutes.

THERE IS ANOTHER PHILOSOPHICAL TOPIC which has so far been mentioned only in passing but which must have come up in H3: probability. Most of the dons there believed that probability presented problems that could not be solved merely through linguistic disentanglement.

Thinking about probability was one of Popper's favorite modes of relaxation: he would cover pages and pages with scrawled equations. There was a link here with his criterion of falsifiability. Quantum mechanics, which deals in probability, was a relatively new branch of physics. It states that the movement of individual electrons cannot be predicted precisely, but only with a degree of probability. Clearly Popper did not wish to dismiss such statements as illegitimate, but how could he absorb probability within his theory of falsifiability? If I say, "The probability of G. E. Moore attending the MSC meeting is only one in ten," then it appears that my hypothesis will not be falsified regardless of whether Moore appears or not. Even if Moore does come, I have not been disproved. I did not say he definitely would not appear: I merely said it was unlikely.

Probability preoccupied not just Popper, but Broad, Braithwaite, Wisdom, Waismann, Schlick, Carnap, and John Maynard Keynes. Unlike many abstruse areas of philosophy, it is a concept we all understand and manipulate in everyday life. Indeed for some people, such as those employed in the insurance industry, it is nothing less than their livelihood.

What are the odds of Red Rum winning the Grand National? What is the probability of a die landing on the number six? What is the chance of a male smoker living into his eighties? What is the likelihood of a nuclear holocaust before the year 2050? Despite the familiar questions, there are few topics more baffling than that of accounting for probability. A fundamental question is whether we talk about probability because it is an objective constituent of the world or only because we are ignorant of what is going to happen. In other words, is the future intrinsically uncertain, or is uncertainty simply the product of our human limitations? In his first book, *A Treatise on Probability*, Keynes inclined to the latter position. Believing that economics, and much else, could be illuminated by an understanding of probability, he maintained that it makes sense to view probability against a background of evidence. At the bookie's, if the only information you have about two competing sprinters is that one is twenty-five and the other fifty-five, it would seem sensible to place your pay packet on the younger of the two. But if you then discover that the man in his twenties is a beer-guzzling smoker and horribly unfit, whereas the older man is a former Olympic gold medalist who follows a strict vitamin-enhanced diet and pumps iron daily in his local gym, you would be wise to adjust your assessment of the odds. The contestants have not changed, but your knowledge about them has.

Others, however, argued that a statement of the form "The probability of throwing a coin three times and getting three heads is one in eight" is simply an *a priori* statistical or mathematical truth—something logically independent of experience, like 2+2=4. One implication of this is that such statements would then not be susceptible to revision on account of new evidence. If a die repeatedly landed on the number six, it would suggest only that it was loaded: it would not undermine the truth of the *a priori* proposition "The probability of a die landing on the six is one sixth."

A drawback of this attitude is that it does not help us with dice in the real world. Insisting on this general mathematical truth is no good to us if what we are concerned about is filling our wallets at the casino by betting on the dice thrown on the craps table. Some in the Vienna Circle therefore championed the "frequency interpretation" of probability, according to which the statement "The probability that this die will land on a six is a half" means only that, if there were an infinite number of throws with this die, then a six would come up fifty percent of the time. But this frequency interpretation of probability is hardly satisfactory: we want to know what the probability is of a six on the next throw of this die, not what will happen in an infinite sequence of throws.

Probability was one of those issues to which Popper kept returning. On trips to the UK in 1935–36 he lectured on it. And on his grant-application form to the Academic Assistance Council he described himself as a specialist on the topic. Throughout his life, his concern was to combat the subjectivism inherent in Heisenberg's uncertainty principle and what is called the Copenhagen interpretation of quantum mechanics. This is subjective in that it states that there are some things we necessarily cannot know about the world: we can never record the movement of atomic particles with absolute precision. We can define the position or the momentum of a particle, but not both at the same time. We can deal only in probabilities. This disturbed not just Popper, but Albert Einstein too. God, said Einstein, does not play dice. He insisted that the world was fully determinate, that it followed the normal rules of cause and effect and that in theory one should be able to predict the trajectory of a particle with 100 percent certainty. To the end of his life, Einstein sought a complete theory that did away with uncertainty.

Popper resolved the dilemma between his objectivist intuitions and Heisenberg's uncertainty principle in another way. He argued that, yes, probability exists in the world, but, no, that does not mean that the world is subjective. It is not as a result of our ignorance that we talk about probability. It is rather that propensity (Popper's preferred term) exists in nature itself. It is an objective component of the world. It is an actual physical reality, akin to an electric force. To put it another way, there is a certainty about probability.

As far as falsification is concerned, he thought that statements involving stable propensities—such as "The die has a one in six chance of landing on six"—could be tested by looking at what happens in the long run. But isolated statements of propensity—such as "There is a propensity of $\frac{1}{100}$ that there will be a nuclear holocaust before the year 2050"—may resist testing, and to that extent exclude themselves from science. One can repeatedly test the tossing of a coin, or the chance of having twins, but not the likelihood of nuclear Armageddon.

ANOTHER PROBLEM WITH an even longer pedigree than induction or probability also surfaced in H3: How can we make sense of the idea of the infinite?

Here was a question which went back to the ancient Greeks. In the fifth century B.C., Zeno of Elea had devised some ingenious brainteasers involving the idea of infinity. Zeno believed that motion and time, as ordinarily understood, were illusory. He thought he had proved either that movement is not possible at all or else that it requires an infinite amount of time.

Two of Zeno's paradoxes involve races and tracks. Zeno argued that an athlete can never run around a stadium, because first he will have to run half the distance, then half the remaining distance, then half the remaining distance, and so on. To complete the circuit he must complete ½ of the total circuit, then another ¼, then another ⅛, 1/16, 1/32, 1/64, and so on. The fractions remaining become closer and closer to zero, but never reach it; the sequence is infinite. Indeed, by the same logic, the ill-fated runner could not make any progress at all, because to move to any spot beyond the starting line he would first have to reach the spot halfway toward this spot, which could only be reached if he made it to the quarter spot, the eighth spot, the sixteenth spot, and so on. The athlete is logically condemned to remain in the starting blocks.

The most famous of Zeno's paradoxes concerns a race involving two competitors: the almost invulnerable Greek hero Achilles and a tortoise. The slow-moving tortoise is given a head start. According to Zeno, speedy Achilles can never overtake the reptile: when he reaches the point at which the tortoise began, the tortoise will have moved forward to a new position, and when Achilles reaches this point, the tortoise will have moved a little farther, and so on.

Many of Zeno's paradoxes are debated to this day. Aristotle's discussion of them helped ensure their survival, introducing a distinction between "actual" and "potential" infinity. Aristotle argued that we can make sense only of "potential" infinity. Thus, for example, the distance around a track is infinitely divisible in the sense that, however many parts it has been divided into, it can always theoretically be divided into more, but not in the sense that it can ever really be divided into infinitely many parts—that is, it always has a "potential" infinity of parts, but never an "actual" infinity of parts.

For over two millennia this was the orthodox dichotomy, the framework within which the concept of infinity was understood. It was not until the arrival of the German mathematician Georg Cantor, writing in the second half of the nineteenth century, that mathematicians found a way of taming infinity, of expressing it in terms that were graspable.

Cantor, referring back to Aristotle's distinction, argued that infinity has an actual, not merely a potential, existence. He described two infinite sets as being equal in size where there is a one-to-one pairing of their members. So, for example, the infinite set, 1, 2, 3, 4, 5, . . . is equal in size to the set 1, 5, 10, 15, 20, . . . because 1 can be paired off with 1, 2 with 5, 3 with 10, and so on. Through such one-to-one correspondence, some of the complications and mysteries of infinity can be unlocked. In particular, Cantor believed himself to have shown that a rigorous mathematical treatment of the actual infinite is possible.

But this approach threw up paradoxes of its own, one of which was exposed by Bertrand Russell, who used as an example Laurence Sterne's novel *The Life and Opinions of Tristram Shandy, Gentleman*. In the novel, Shandy spends two years writing up the first two days of his life. He worries that at that rate he will never finish his autobiography. Russell argued that applying Cantor's mathematics, strangely, if Shandy lived forever there would be no day that would go unchronicled. If from his twentieth birthday he spent two years working on his first two days, then from the age of twenty-two he could cover the next two days, from the age of twenty-four the next two, and so on. Although he would fall further and further behind, there would still be a one-to-one relation: each day of his life has a corresponding period of autobiographical activity—

Age 20–21	Days 1–2
Age 22–23	Days 3–4
Age 24–25	Days 5–6

An everlasting Tristram Shandy could apparently write up every day of his life.

In 1946 the issue of whether there are both "actual" and "potential" infinities was very much alive, and it is known that it arose in H3.

20

Slum Landlords and Pet Aversions

> *I remember, after one particularly fatuous paper at the Moral Sciences Club, Wittgenstein exclaiming, "This sort of thing has got to be stopped. Bad philosophers are like slum landlords. It's my job to put them out of business."*
>
> —MAURICE O'CONNOR DRURY

IN *Unended Quest*, Popper makes plain his attitude to the Cambridge meeting: "I admit that I went to Cambridge hoping to provoke Wittgenstein into defending the view that there are no genuine philosophical problems, and to fight him on this issue." The thesis that there are no genuine problems, only linguistic puzzles, was among his "pet aversions."

Exactly how the philosophical row proceeded in H3 we cannot be sure. But clues lie in the minutes, in Popper's account, in eyewitness reports, and in the deferential letter Popper wrote to Russell the day after he returned to London.

There were real problems, Popper maintained at the meeting, not just puzzles. Breaking in, Wittgenstein "spoke at length about puzzles and the non-existence of problems." Popper describes himself as interrupting in turn with a list of problems he had prepared. The existence of actual or potential infinity, induction, and causation were all brought up. The problem of infinity was dismissed by Wittgenstein as being merely a mathematical question, no matter whether Cantor had or had not satisfied mathematicians with his method of dealing with the infinite. Induction "Wittgenstein dismissed as being logical rather than philosophical."

At some stage during the argument, Russell weighed in on Popper's side and cited the British empiricist John Locke. This may have been in reference to what Locke has to say about the question of personal identity, and what it is that makes me the same person now as I was thirty years ago. (Locke's answer was that it is the continuity of the mind and memory.) It may have been Locke's distinction between primary and secondary qualities—the distinction between, for example, shape and color. (Locke maintained that primary qualities exist, as it were, in the objects themselves, whereas secondary qualities are parasitic upon the observer. A square is still a square when unobserved, but a red square depends for its redness upon the existence of beings whose perceptual apparatus causes them to see it as red. Secondary qualities, unlike primary qualities, cannot be understood without reference to the conscious mind.) However, it is most likely to have been Locke's claim that there is no such thing as innate knowledge—that all our knowledge comes from experience, that it is out of experience that the mind fashions its ideas, and that it is only our mental ideas (or, in Russell's terms, sense data) to which we have direct access. If that is so, then the problem remains of how we can have secure knowledge about anything outside our minds—how we can know about other minds and other things.

In any case, Popper found the Locke interjection helpful and wrote to Russell to say so. And his letter goes on to detail the real substance of his paper, though the fact that he needed to spell it out implies that he had failed to put his arguments beyond doubt—as he always demanded of others—the first time around.

The kernel of Popper's critique was this. If Wittgenstein wants to reject out of court a question of the form "Can something be both red and green all over?" then he needs to explain on what grounds. To differentiate propositions that are acceptable from those that are not, some sort of theory of meaning is required. And this must be a problem, not a puzzle.

Wittgenstein's claim that there are only puzzles is itself a philosophical claim, Popper avers. This claim may be correct, but Wittgenstein has to prove his case, not assert it. And in endeavoring to prove it he will necessarily be sucked into a debate about a real problem—the problem of justifying the exact position of his frontier between sense and nonsense, justifying why some things can legitimately be said, and other things cannot, why some sentences have meaning and why others are meaningless. So, even if most philosophy is about puzzles rather than problems, there must be at least one problem, even if all other apparent problems are merely puzzles.

Wittgenstein had foreseen this objection, but his response was to remain mute. Just as in the *Tractatus* the pictorial relationship between language and the world could not itself be pictured, so to try to mark the boundary between sense and nonsense was to trespass over this very same boundary. "Whereof one cannot speak, thereof one must remain silent."

21

Poker Plus

Let's cut out the transcendental twaddle when the whole thing is as plain as a sock on the jaw.

— WITTGENSTEIN

TAKE A DISPUTE FUNDAMENTAL TO philosophy, for whose future both men felt personal responsibility; take the cultural, social, and political differences between the protagonists; take the obsession of one with the other, who is in turn totally self-absorbed; take their no-holds-barred style of communication; take their complex relationship with their father figure, Russell—throw all these into the cauldron that was H3 and a major explosion seems to have been inevitable. The poker becomes only a fuse. That much is certain. We must also constantly remind ourselves that both protagonists were exceptional—but one was all too human, one not quite human.

The questions remain: Was Popper's account of the meeting wrong? Did he lie?

In constructing a narrative for that evening, we can be reasonably certain of some facts: for instance, about the Cambridge through which the participants walked to the meeting.

The autumn evening was unusually cold, gripped by that damp chill which makes those whose joints have ached there name Cambridge "the University of the Fens." Even to the ruddy-faced college sportsmen it felt the colder because Britain, though victorious in the war, was still living under wartime scarcity.

The streets, lecture theaters, and college courts were thronged with the not-long demobilized: the twenty-three-year-old former captain who had scrambled up the beaches of Normandy or sweated through the jungles of Burma; the former air gunner who still bore the wan complexion of his prisoner-of-war camp; the one-time naval lieutenant who had spent four years in destroyers convoying food and oil; the former Bevin Boy with memories of the heat and dirt of the coal mines—these were now the serious-minded undergraduates whose main ambition was a "war degree" in the two years permitted and then to get on with the business of life. Among them, the smooth-chinned youngsters straight from school stood out, uncertain whether to be glad or sorry that they had missed the big show. It was all too easy to join a conversation about the disaster at Knightsbridge Corner only to discover that the subject was not some accident near Harrods but a battle in the desert war.

Students showed little joie de vivre. Their daily lives were governed by shortages—worse than during the fighting, they grumbled. Even bread was now rationed (as it had not been in the war, for fear of riots), and fuel was in short supply. Trinity, with its vast country estates, regularly offered jugged hare and venison. Less well-endowed colleges served up Poor-Man's Stew—made out of bones—while dons luxuriated in pigeon pie. Perpetually hungry undergraduates rose at dawn to queue for rolls and cakes brought up from London and sold in the market square, the ex-servicemen and -women looking back with unexpected nostalgia on mess or wardroom, canteen and Naafi. The health-food shop in Rose Crescent regularly ran out of nut cutlets. Celebratory college feasts merely left the celebrants setting off to look for another meal. When King's held a feast the next year, the college history records, "The combination of restricted food and unrestricted Algerian wine produced something of a shambles."

On 25 October 1946, for those with little interest in abstruse logical problems (or puzzles), alternative entertainment was on offer. Politically minded undergraduates could go to the University Labour Club to hear the achievements of the new Labour government charted by the sonorous Minister for National Insurance, James Griffiths, a son of the Welsh valleys and former miner; they would have time for half a pint of thin beer afterward, and still be back in college before the gates closed at eleven—not that climbing in was much of a challenge.

On the BBC Light Programme, Victor Sylvester's ballroom orchestra was playing strict tempo for his "Dancing Club." On the more serious-minded Home Service, there was a discussion on the nationalization of the electricity industry—part of the government's commitment to a planned economy, taking basic industries and services such as railways, mines, and aviation into public ownership. And listeners to the recently introduced Third Pro-

gramme, dedicated to high culture, could savor extracts from
Chaucer's poem *The Canterbury Tales*, followed by contemporary
French prose from Paris. For the connoisseur of classical music
prepared to make the journey to London, in the Albert Hall Popper's distant relative and a former grateful visitor to the Palais
Wittgenstein, Bruno Walter, was conducting Mozart's G minor
symphony to the music critics' unreserved approval.

The newspapers, as thin as during the war, were chewing over
how Goering, sentenced to death by the International War
Crimes Tribunal as a leading Nazi war criminal, had escaped
hanging by swallowing a poison pill—mysteriously defying the
constant searches of his Nuremberg prison cell. Also in Germany,
the first signs of East-West division were appearing. The Ameri-
cans were anxious to revive the German economy, and there were
tensions in Berlin. At the United Nations, then located in Flush-
ing Meadow, Queens, New York, members were debating nuclear
energy. Two of England's cricketing heroes, Hutton and Wash-
brook, gave some good cheer to their exhausted compatriots with
an unbroken opening stand of 237 against South Australia. . . .
Tired out maybe, but England could still show them. *The Times*'s
personal column hinted that some of its readers were feeling the
pinch. For sale: full-length fur coat, army officer's two uniforms, a
1933 Rolls-Royce, a gold watch. Another appeal spoke of an unchanging England: a vicar (with either an extended family or an
extensive social life) requires a living with a large house—bishop's
reference available.

So far, so certain. And yet for the members of the Moral Science Club the evening belonged to two exiles from Vienna whose
lives could so easily have intersected within the Ringstrasse ten or
twenty years before. Their individual paths had finally brought

them face-to-face in the most established of English academic surroundings.

In attempting to piece together what happened that night, we must understand that Popper and Wittgenstein came to the meeting in quite different states of mind and with quite different objectives. For Popper, combat and a culminating moment beckoned. For Wittgenstein, a chore, an obligation to be fulfilled: keeping both the MSC and philosophy free from the contagion of problems.

It was ten years after Popper had first spoken at the Moral Science Club, when Wittgenstein had been absent with a cold. But this was a visit with a difference. In 1936 Popper had been a dislocated figure, supported by his wife's schoolteaching in Austria while he searched for the permanent university post for which his Jewish background would have been a barrier in Vienna if not a disqualification. He had been short of money, "oversensitive" to his lack of success, and staying in squalid quarters. A decade later he was set up for the future: he had a secure position, a confident and independent philosophical voice, and, at last, recognition where it mattered — in Britain. He had arrived from New Zealand to be greeted by respect and admiration for *The Open Society*, finally published in London in November 1945. At forty-three, it had been a near thing: he had feared that no English university would want to import a lecturer over forty-five.

Reviewing *The Open Society* in the *Sunday Times*, the political scientist and classical scholar Sir Ernest Barker saluted "an abundance of riches — classical scholarship, scientific acumen, logical subtlety, philosophical sweep." The historian Hugh Trevor-Roper described it as "a magnificent and timely achievement . . . by far the most important work of contemporary sociology. . . . [Popper] has restored significance to human choice and the human will."

Not all the critics were so enthusiastic. The anonymous reviewer in the *Times Literary Supplement* (it was Harold Stannard of *The Times*) led the paper under the headline "Plato Indicted:" "Dr. Popper's book is a product of its time; and as the time is earnest, critical and aspiring, so is the book. In its strength and its weakness, its sincerity and its dogmatism, its searching criticism and its intellectual arrogance, it is typical or, at any rate, symptomatic of the age." Six months after the H3 meeting, in April 1947, Gilbert Ryle, reviewing *The Open Society* for *Mind*, took the same line. On the one hand, he praised "a powerful and important book. It is a criticism of a set of dogmas which underlie the most influential political theories and in consequence powerfully affect the actual conduct of human affairs." But, on the other, he had serious reservations about the author's tone, fretting that Popper risked "diversion" by his "vehement and sometimes venomous" strictures, that his comments had a "shrillness which detracts from their force. . . . It is bad tactics in a champion of the freedom of thought to use the blackguarding idioms characteristic of its enemies." Though Wittgenstein claimed never to read *Mind*, he knew about Ryle's review and was disgusted by it. The reason was almost certainly the tip that Ryle had for his readers: "Don't miss the notes which contain interesting and important *aperçus* on the esotericisms of Wittgenstein." ("Esoteric" was a term Popper used in an extended note in *The Open Society* severely criticizing Wittgenstein.)

"Vehement," "venomous," "shrill:" would those be appropriate adjectives to describe Popper's tone in H3? Certainly his target for that night was what he saw as Wittgenstein's destructive influence on philosophy. Perhaps, too, he had a more pressing score to set-tle, convinced that Cambridge University Press, the first British publisher approached, had turned down *The Open Society* to pro-tect Wittgenstein. Although CUP did not generally give reasons for rejecting a book, von Hayek was told in confidence that with *The Open Society* there were two. He passed them on to Gom-brich, who in turn sent them to Popper in New Zealand. Its length was against it, but also a university press ought not to pub-lish something so disrespectful of Plato. On hearing this, Popper commented, "I still suspect that 'Plato' is only a euphemism for the three Ws: Whitehead, Wittgenstein, Wisdom."

There was one other Cambridge figure that Popper had in his sights that night: Russell. His claim to be Russell's intellectual heir and his patent anxiety to impress Russell form a subplot to the H3 confrontation.

For Wittgenstein, this was an MSC meeting like any other in the past thirty-five years. But, to add to the prospects for a major clash, he went to H3 in a black humor and in the throes of an in-tense loathing of Cambridge. A month earlier he had recorded, "Everything about the place repels me. The stiffness, the artificial-ity, the self-satisfaction of the people. The university atmosphere nauseates me." He constantly considered giving up his chair.

He was also exhausted. That term he devoted a great deal of time to students: twice-weekly classes of two hours each, a weekly at-home of two hours, a whole afternoon spent with Norman Mal-colm, another spent with Elizabeth Anscombe and Wasfi Hijab. A proselytizer for vitamins, Wittgenstein had discovered vitamin B, used to combat tiredness and mood swings. But, with or without vitamins, teaching always left him in a state of nervous fatigue.

Was his adversary on his mind? Probably not at all. Before this date, Wittgenstein had appeared oblivious of his fellow Viennese philosopher and his determination to lock horns. A couple of weeks earlier, when Peter Munz mentioned that he had studied under Popper in New Zealand, Wittgenstein had replied, "Popper? Never heard of him." That this was true is a distinct possibility, given Popper's recent obscurity and Wittgenstein's lack of interest in contemporary philosophers.

In any case, Wittgenstein's notebooks for the period reveal quite different philosophical concerns—for instance with the complex grammar of color words—and deep personal preoccupations. On the personal side there was Ben Richards, a medical student on whom Wittgenstein became fixated. When Wittgenstein resigned his chair the following year and moved to Ireland, Richards would go and visit him. Wittgenstein urged Richards to read American crime stories. On the day of the meeting, using a code—A=Z, B=Y, C=X, etc.—which he had learned as a child and could write as fluently as normal German, he scribbled, "B has a thing about me. Something that can't last . . . whether it will work out . . . I do not know, nor whether I can endure this pain. Demons have woven these bonds and hold it in their hands. They can break them or they can let them survive."

It is improbable that Richards did have "a thing" about him. Wittgenstein was apt to imagine that relationships were more meaningful than were often the case. There is no other evidence that Richards (who later married) was homosexual. In any event, on the day after the MSC meeting, on 26 October, Wittgenstein continues in the same vein, musing over the value of love:

> love is THE [sic] pearl of great worth which you hold to your heart and which you will never exchange for anything and which one

can reckon to be the most valuable thing. It shows us moreover if
one possesses her (love), what great value is *[sic]*. You get to know
what it means, to recognize its value. You learn what it means to
extract precious stones.

It was in this state that he left for H3. And at this point, with that
background in mind, let us reconstruct the probable sequence of
events, before turning to weigh the evidence.

ALONE AT LAST! Ludwig Wittgenstein finished the tomato
sandwiches he had bought from Woolworth's earlier that day. He
ate them in memory of Francis, whose favorite they had been,
rather than to satisfy his hunger, and left his room. Outside, the
little landing and the steep wooden stairs reminded him of the
maids' quarters in the Alleegasse house, except that those folding
chairs would never have been stacked so untidily there—or
stacked at all. The upper servants would simply not have permitted it. He restacked the chairs—deck chairs with deck chairs, garden chairs with garden chairs—making sure that each was exactly
under the one above, the stacks symmetrically placed in orderly
rows. Was the memory of tomato and bread or of Francis? Why
did the memory of the one trigger the memory of the other?
Were memories piled one on another like the chairs? And now
the MSC. . . .

Intolerable . . . Intolerable . . . And how could he think
clearly when Ben was on his mind the whole time. Ben . . . What
hope had he that Ben would feel the same way? Still, he'd share
the latest Max Latin with him. And the meeting—he would give
it an hour and a half. He strode into Trinity Street and turned left.

Something made an undergraduate pause in the entrance to

Caius. What was it about that man? The evidently military step? The cropped graying hair? The neatness? A keen birdlike gaze? A senior officer paying a visit, or returning to his fellowship? Whatever the case, a fleeting impression of an unusual intensity lingered.

In H3 the coal fire, the only source of warmth, had been banked up and produced a dismal heat. Braithwaite picked up the poker and cleared out some of the ash in the hope of making the fire draw better. His efforts were rewarded with an indecisive plume of smoke that died away as he watched. The blackout curtains were grimy and torn, adding to the room's long-undecorated drabness. Braithwaite turned back to his guest with a question, but that too died in the air as Popper, absorbed in his notes, muttered to himself in German.

The members who packed the room—many more than the available chairs—were indifferent to their surroundings. The atmosphere was expectant. Dr. Popper's newly published book was something of a cause célèbre. A Girton don had forbidden her students to read it as it was too scandalous in its attack on Plato. Communists and Labour left-wingers were up in arms, too, but over its attack on Marxism and planned societies. The speaker was Viennese, like the club's chairman, Professor Wittgenstein, but was understood to be flatly opposed to the language-based approach of his fellow Austrian. Braithwaite, who knew Popper, had predicted fireworks, a fraught occasion. The word had spread: here at last was someone who could take on Wittgenstein, who would not be crushed beneath the juggernaut. Hadn't Popper—the only one of its members not to be in thrall to Wittgenstein—destroyed the Vienna Circle with a single devastating insight?

And then still only in his early thirties. And, truth to tell, the prospect of a clash had a certain appeal for those who crowded in, bored with the MSC's usual fare of worthy meetings dominated by the insistent monologues of one man. And from that point of view, the very opening words uttered by the guest were full of promise.

Popper couldn't wait to begin. His energies were surging, his heart was banging with the extra adrenaline. He pulled his earlobe down, partly to listen to the chatter, to see what points were being anticipated, partly to calm himself. This was the moment, and he was the man. Recognition was his at last in the greatest country in the world: *The Open Society* had transformed political philosophy, just as *Logik der Forschung* had clarified once and for all the method of science. Invitations to speak were pouring in. The LSE was just a beginning. And tonight he would achieve a third triumph. He would dispatch the rubbishy notion that playing with words was philosophy, dispatch this *Scharfmacher* with his impossible *Wichtigtuerei*—this self-important agitator. And Russell, yes Russell, was on his side, had urged him on—in Newton's chamber, what's more—vanquishing any doubts that he had chosen the right issue for the battle they both wanted. How could anything be more apt than for the man who had formulated the falsifiability test to sit in the room of the scientist whose laws had had God-given status but were now falsified? And to sit there with the greatest thinker since Kant! Tonight he would win. And Wittgenstein apologize.

And could one ask for a greater victory? Wittgenstein exposed. The exalted brought down. The inspiration of the Vienna Circle, always carefully keeping himself apart; the lonely genius prowling the Halls of Wittgenstein. It had even become a coffeehouse joke that Wittgenstein didn't exist—he was a figment of poor Schlick's and Waismann's imagination, their golden mountain. Tonight the world would discover how real he was. . . . Popper looked round the audience. Ewing was staring at his boots; Wisdom reading some racing paper no doubt. Braithwaite smiled encouragingly. His wife was uncrossing her legs. A foreign-looking student shifted uneasily in his seat.

When the guest opened the meeting, there was no question of the normal courtesies. A former naval officer in the audience recalled Admiral Fisher's maxim "Hit first, hit hard and keep on hitting" as Popper went straight in with a frontal assault against the wording of the invitation: to deliver "a short paper, or a few opening remarks, stating some philosophical puzzle." Whoever had written the reference to puzzles had, perhaps unwittingly (said with a slight smile), taken sides.

This was a comment the guest felt he had made in a suitably lighthearted manner. But for one person present it was more of a challenge than a piece of lightheartedness: the gauntlet was picked up.

Intolerable. This was intolerable. Wittgenstein wouldn't allow it. Why listen to such foolishness from this upstart, this *Emporkömmling*, about a formal invitation for which the secretary wasn't even responsible? The wording was his. The point of it was to cut the twaddle and get down to business. Wittgenstein sprang to the secretary's, his student's, defense. Loudly. Insistently. And, felt Popper, angrily. The evening had started as it was to go on.

Grateful for his defender's immediate and ferociously direct counterattack, the secretary, Wasfi Hijab, scribbled furiously, try-

ing to keep up with the quick-fire interchanges, the voices rising
and falling over each other like angry seas running on to a beach:

> *Popr:* Wittgenstein and school never venture beyond
> preliminaries, for which they claimed the title philosophy, to
> the more important problems of philosophy . . . gave some
> examples of difficulties whose resolution required delving
> beneath the surface of language.

> *Wittgen'n:* these are no more than problems in pure math'ics or
> sociol'gy.

> *Aud'ce:* unconvinced by Popper's examples. Atmos charged.
> Unusual degree of controversy. Some very vocal.

(The thought flitted through Hijab's mind that the minutes would
be more fun to write up than usual. He would do them tomorrow.)

But now, by a sort of reflex, Wittgenstein's hand had gone to
the hearth and tightened around the poker, its tip surrounded by
ash and tiny cinders, as Braithwaite had left it earlier. The don
watched anxiously as Wittgenstein picked it up and began con-
vulsively jabbing with it to punctuate his statements. Braithwaite
had seen him do it before. This time Wittgenstein seemed especially agitated, even physically uncomfortable—unaccustomed to
a guest's counterpunching, perhaps. By this stage of a meeting he
was usually in the full flood that people complained about behind
his back. Braithwaite suddenly felt uneasy: should he try for the
poker? Things were beginning to look somewhat out of control.

Someone—was it Russell?—said, "Wittgenstein, put the poker
down."

Wittgenstein was conscious of pain, a constant distress, as if he were listening to a gramophone record playing at just the wrong speed. This mushy thinking! It was bad enough that this ass, this Ringstrasse academic, was expounding a theory, was trying to say things which couldn't be said, was deluding himself into believing that there were hidden depths into which he could delve—like a man who insisted on digging an underground shaft in an open-cast mine. . . . In itself this was bad enough. But not even attempting to open his mind to clearing out this rubbish, not to listen to what he himself was saying . . . This had to be stopped, the malignancy cut out.

Somewhere in the back of his mind Popper knew he was going too far. Tomorrow he would feel remorse for failing to control himself, just as after the Gomperz evening in Vienna—though he had never managed to admit that to poor Schlick. This Wittgenstein was real enough. But who would have said "mystic?" All the dogmatism of a Jesuit. And the fury of a Nazi. A maniac misleading philosophy—he had to confess he was completely wrong. Just one more push, one more brick knocked out of this tower of chitchat. And now the madman had picked up the poker and was jabbing away as he tried to interrupt. Jab, jab, jab, in time with his syllables. "Popper, you are WRONG." Jab, jab . . . "WRONG!"

Unattended, the fire was almost out. It was no matter: being at the meeting was now like being trapped in a hothouse and entangled in jungle creepers. With the clash of angry voices, the running interjections from Wittgenstein's disciples, the unprecedented crowd—those standing (the "wallflowers") pressing in not to miss a blow being struck—the audience was caught in a blinding confusion. A literary-minded undergraduate took refuge in Matthew Arnold:

> . . . a darkling plain
> Swept with confused alarms of struggle and flight,
> Where ignorant armies clash by night.

He wondered again if he shouldn't change to English for Part II.

Hang on! "Flight" was right, for Wittgenstein had thrown down the poker and was now on his feet. So was Russell. In a sudden moment of quiet, Wittgenstein was speaking to him.

"You always misunderstand me, Russell." There was an almost guttural sound to "Hrussell."

Russell's voice was more high-pitched than usual. "No, Wittgenstein, you're the one mixing things up. You always mix things up."

The door slammed behind Wittgenstein.

Popper stared disbelievingly at Wittgenstein's empty chair. Russell was saying something about Locke. Had he won? Driven Wittgenstein out? Left him with nothing to say? Killed him off, like the Vienna Circle? But where was the confession that he was wrong? The apology? Someone was addressing him. It was his host for the evening, Braithwaite, asking in his kindly fashion for an example of a moral principle. A picture of the poker came into his mind. "Not to threaten visiting lecturers with pokers." There was a pause and some laughter—rather like that time before the war, when the audience had thought, wrongly, he was joking. Well, he'd shown them.

The questions started again, but this time typically understated English questions. He answered them almost absentmindedly. Had he won? Someone—a Wittgenstein supporter apparently— put a question designed to catch him out: could Sir Henry

Cavendish's experiments be described as science, given that they were conducted in secrecy? "No." He shut him up and returned to savoring his battle with Wittgenstein. Russell would agree he had won. Wouldn't he?

Out in the deserted street with the huge bulk of the silent chapel looming over him, Wittgenstein was taking deep gulps of cold air. He began to think about a puzzle that had come up in his seminar that afternoon: in comics, a balloon with words means "speaking," a cloud with words means "thinking." What does this tell us? In a room above a shop in King's Parade, an undergraduate had tuned his radio to the Third Programme. Through the open window Dylan Thomas could be heard—that light Welsh accent, rounded vowels, almost singing:

Wittgenstein and Ben Richards.
"B has a thing about me.
Something that can't last."

August Bank Holiday. A tune on an ice-cream cornet. A slap of sea and a tickle of sand. A fanfare of sunshades opening. A wince and a whinny of bathers dancing into deceptive water. A tuck of dresses. A rolling of trousers. A compromise of paddlers. A sunburn of girls and a lark of boys. A silent hullaballoo of balloons. . . .

Karl and Hennie. He wrote and wrote. She typed and typed.

22

Clearing up the Muddle

As most lawyers know, eyewitnesses often err. . . . If an event suggests some tempting interpretation, then this interpretation, more often than not, is allowed to distort what has actually been seen.

—POPPER

"Now this was a case in which you were given the result . . . now let me endeavour to show you the different steps in my reasoning."

—SHERLOCK HOLMES IN SIR ARTHUR CONAN DOYLE'S
"A STUDY IN SCARLET"

UNDOUBTEDLY, IF THE JOKE had triggered the walkout, it would have been a heroic feat for Popper. Unusually for him, there seems to have been a personal element in the intellectual duel. He takes aim and fires. He scores a palpable hit. The wounded duelist leaves the field. Leaves it to Popper and his second, Russell.

But on a balance of probabilities it seems doubtful that it was Wittgenstein who asked Popper for an example of a moral rule. Peter Geach and the late Casimir Lewy, a Polish-born specialist in philosophical logic, who both called Popper a liar over his version, may have justice on their side, if not professional courtesy. Even those such as Sir John Vinelott, who first asserted that it was Wittgenstein who had posed the question, later admitted their doubts.

It seems likely that so dramatic an incident—the chairman puts a question and is so discomfited or angered by the quip made in answer that he throws down a poker and walks out—would have found a place in the minutes. And another, closer, look at *Unended Quest* also casts doubt on Popper's account. Popper describes himself as putting forward, one after another, the list of problems he had prepared. Wittgenstein brushes these aside, pounding on about puzzles and the nonexistence of problems. But Popper does not record Wittgenstein's asking a question—until, that is, he suddenly demands a moral principle. This comes out of nowhere, quite at odds with the run of their dialogue.

As was his habit, Wittgenstein had certainly interrupted continually, attacking the various illustrations Popper gave of philosophical problems—induction, the question of whether we can know things through our senses, the existence or otherwise of potential or even actual infinities. But the utterance of the poker principle seems more of a piece with the probing interchanges between Popper and Wittgenstein's disciples that went on after their master had quit the battle. Peter Geach, for instance, tried to trap Popper by asking whether experiments carried out by Sir Henry Cavendish would be correctly described as science. Cavendish, who is most famous as the discoverer of hydrogen and other gases, was so secretive a researcher that he put a second staircase in his home for the servants to avoid meeting them. He is said to have spoken fewer words in his life than a Trappist monk. Popper insisted that a theory could be validly described as scientific only if it were both falsifiable and open to scrutiny. To Geach's question, therefore, Popper said simply, "No."

Wittgenstein's reported exchange with Russell suggests the real catalyst for this evening's premature departure. If anybody could touch Wittgenstein personally, Russell could. Wittgenstein being Wittgenstein, there was no question of his staying put for politeness's sake. And this week he had been deprived of his usual monologue. That much, at least, was due to Popper.

As for the suddenness of the departure, Popper, of course, could not have known that, despite being in the chair, Wittgenstein routinely left the MSC early even when in a calmer frame of mind. He always walked briskly, in military fashion, and, according to Peter Munz, he never shut a door quietly. The previous term, when A. J. Ayer had addressed the MSC, Wittgenstein had withdrawn before the end and without exchanging a single remark with the guest. Ayer described Wittgenstein's departure as "noisy." To the tense visitor on 25 October it must have seemed that Wittgenstein had stormed out.

However, the accusation leveled against Popper was not simply that he was mistaken. It was that he—a leading philosopher, the destroyer of Plato and Marx, admired by presidents, chancellors and prime ministers—had lied in his autobiography, in an account so definite and logical.

The fireplace. The poker dropped on the tiles of the hearth "with a little rattle."

A lie? Although Peter Geach has made that accusation in the past, he is now inclined to give this episode a gently indulgent interpretation. He quotes Shakespeare's Henry V imagining a veteran of Agincourt:

> Old men forget; yet all shall be forgot,
> But he'll remember with advantages,
> What feats he did that day. . . .

But this is far less generous an exculpation than it might seem at first sight. To claim that you had hit back at your opponent to his face when it was really behind his back after he had left the field smacks more of Falstaff than of Henry's truly valiant soldiers. In fact a glance at Henry's speech shows that Geach risks quoting against himself. Henry is predicting that the veteran of Agincourt will remember so vital an actual encounter, albeit with some embroidery ("advantages") in the telling of the deeds he performed in it: he'll remember that, whatever else he might forget. And if it is the phrase "Old men forget" that expresses Professor Geach's latter-day indulgence, he does Popper an injustice. When Popper composed his autobiography, written in the first instance for Schilpp's *The Library of Living Philosophers*, he was in his mid-sixties. But he had not long retired from teaching at the LSE with the aim of writing full-time, and he was still to publish two major works. He was as mentally energetic as ever.

It is significant that there is no comparable narrative error in *Unended Quest*—nothing else to be put down to the lapses in memory of an elderly man. Popper's account of 25 October is not only unusually detailed but is the only extended, vivid anecdote of its kind, bringing to an end the personal side of the autobiography well under halfway through the work. And Popper gave very careful attention to the wording of the story, drafting and redrafting it by hand. He considered, for example, whether he had gone to Cambridge to "incite, seduce, bait, challenge" Wittgenstein. He settled for "provoke."

Popper was well aware that his version of the poker meeting was contentious. His archives contain an undated note—handwritten jottings in German, apparently corrections for a new edition of *Unended Quest*—in which he defends himself against a story circulated by his critics that he was wrong in saying that Russell had been present. There is also a letter written in May 1968, commenting on Professor McLendon's version of the meeting. Popper explicitly confirms his own account of the incident and notes that his memory is "very clear except for the date."

In short, Popper knew what he was doing when he put pen to paper. But was he consciously misrepresenting the events or did he believe his own story? The answer must lie in the nature of his involvement.

Looked at in the context of *Unended Quest* as a whole, the poker story comes across as central to Popper's idea of himself, the outsider who challenges the prevailing view. He defined himself by opposition. At one stage he even planned to open his intellectual autobiography with the poker. If nothing else, this shows that in Popper's eyes the episode was a glorious triumph. Malachi Hacohen believes that Popper saw the incident "as a struggle between giants . . . that he had won." However, in the end he decided that to begin in this way would look like bragging.

There is plainly a streak of self-aggrandizement in Popper, who sees himself as the prime author of events. Autobiographies create heroes of their authors, putting them center stage by definition. But in *Unended Quest*, besides slaying the dragon Wittgenstein, Popper is a hero twice more. He becomes the man who crushed logical positivism: "I fear that I must admit responsibility." And he is the man who helped rescue Friedrich Waismann from Vienna and the Nazis—another example of Popperian embellishment:

> [The New Zealand job] was a normal position, while the hospitality offered by Cambridge was meant for a refugee. Both my wife and I would have preferred to go to Cambridge, but I thought this offer of hospitality might be transferable to somebody else. So I accepted the invitation to New Zealand and asked the AAC and

Cambridge to invite Fritz Waismann, of the Vienna Circle, in my
stead. They agreed to this request.

The implication that Popper turned down the Cambridge opportunity so that Waismann could have it is not borne out by the
letters Popper wrote at the time. Anyway, the temporary lecturership at Cambridge was *ad personam*—it was created specifically
for Popper. And while it is true that Waismann was later to be offered a similar package—a grant from the Academic Assistance
Council and a lecturership at Cambridge—and while it is also
true that Popper strongly recommended Waismann to both Cambridge and the AAC, it was by no means inevitable that, because
Popper chose to turn down the Cambridge post, Waismann or indeed anybody else would be offered the place. Waismann did not
even name Popper as one of his referees.

We should remember, too, the potential to mislead of that fleeting reference in his 1952 lecture to "when I saw him [Wittgenstein]
last," though Popper and Wittgenstein met only once.

Did Popper lie, then? The best guess must be that his imagination had created a fixed—if false—memory. Popper believed
his account to be true.

"Memory is the most paradoxical of the senses," Peter Fenwick, a neuropsychiatrist at the Institute of Psychiatry in London,
has written, "at the same time so powerful that even the most
fleeting impressions can be stored, forgotten completely, and then
reproduced in perfect detail years later, and yet so unreliable that
it can play us completely false." There is a thicket of problems in
assessing claimed memories. Later information, true or false, can
easily distort recollections, even producing conviction about
events that never occurred. The imagined version of the event,

perhaps constantly reimagined, can take over: the subject comes to believe that what he or she has imagined was what happened. The true recollection is wiped out. "How and when false memories are laid down we don't yet know. Some researchers think that they are recorded in the brain at the time of the event; others believe that people develop a schema about what happened and retrospectively fit other events that are untrue, although consistent with their schema, into their memory of the original experience."

Perhaps in Vienna and New Zealand, in lonely contemplation, Popper had imagined such a face-to-face confrontation. Philosophically and personally the prize could not have been higher. He prepared with care, laying out his line of attack, anticipating objections. But there were things he could not have anticipated: such hostility from a bank of students, and, amid that hostility, the poker. The impact the shaking poker might have had on the Cambridge audience is not the measure of its impact on Popper. They were used to Wittgenstein—though that night, even by his standards, he was unusually agitated.

Then Wittgenstein was gone, unexpectedly, and apparently because of something someone else had said. The battle was not over, neither won nor lost; it had simply evaporated.

So heightened and important an occasion can lead to its excited recollection being worked over again and again. The key moments are picked out and dwelt upon. Some of the events are elided; others are shaped into a more satisfactory pattern, another sequence. New causal connections are established. The result of this process becomes fixed; it becomes the memory of the event.

There is also the question of Popper's having given the wrong date for the meeting—26 October—in his autobiography. This is more easily answered. In 1968, when he was asked to check

McLendon's version, Popper appealed to the then secretary of the MSC for the date. She referred to the minutes written up by Hijab on the Saturday and dated 26 October. Had Popper burrowed into the thousands of letters, papers, speeches, and drafts piled up at home in Penn, he would have found the notes he made for his MSC paper—and in them the true date, 25 October.

THERE IS ONE MORE ISSUE to be cleared up, relating to the possible role of Bertrand Russell. Did Russell—at odds with Wittgenstein and thoroughly disapproving of his approach to philosophy—put Popper up to the fight in the cause of saving philosophy from its descent into tea-table patter?

This engaging claim was raised in an article by Ivor Grattan-Guinness. It is based on an interpretation of the letter sent by Popper to Russell in the aftermath of the meeting, and the passages that read:

How much I enjoyed the afternoon with you, and *the opportunity of cooperating with you*, at night, in the battle against Wittgenstein. . . .

My own paper contained only little, as you will remember, I had warned you; it was for this reason that I had considered discussing something else. . . .

Your bringing in Locke helped a great deal. Indeed, the situation is now, I feel, as clear as it can be. . . .

And, after discussing the logical arguments:

This is why I had to choose (and why, *advised by you*, ultimately did choose) this topic. [Italics added.]

But these comments are at best ambiguous. Clearly Popper had spoken to Russell about his paper. But when? Just before the meeting, when they had had tea in Trinity, or earlier?

All this could have been over tea, particularly if Popper did not have a paper as such, only some mental notes. Indeed, perhaps Popper arrived in a state of indecision and was finally persuaded what his topic must be by Russell.

This, however, is scarcely the impression given by *Unended Quest*, nor is it supported by the documentary evidence. Once Popper had fixed the date of his MSC visit, he wrote Russell a letter (not in the archives). Russell replied on 16 October offering to meet Popper on Friday afternoon at four o'clock or on the Saturday morning. The formal tone does not sound as though there had been prior consultation or that the lecture topic was to be discussed—if that was the intention, why give Saturday morning as an alternative?

Then, in Russell's acknowledgment of Popper's postmeeting letter, there is the sentence: "I was entirely on your side throughout, but I did not take a larger part in the debate because you were so fully competent to fight your own battle." Not only does this contain no hint of an earlier collaboration, but the phrase "to fight your own battle" rather suggests that Russell did not see himself as having earlier lined up alongside Popper. "I was entirely on your side throughout" is a strikingly redundant observation if he had put Popper up to the confrontation in the first place.

But it must be admitted that the phrase "(and why, advised by you, ultimately did choose)" remains a mystery. It is, however, curiously close to a phrase in *Unended Quest* on the title of Popper's paper, grumbling that the minutes of the meeting are not quite accurate: "The title of my paper is given there (and it was so given on the printed list of meetings) as '*Methods in Philosophy*' instead of '*Are there Philosophical Problems?*' which was the title ultimately chosen by me." Ultimately—but when? Could it be that the change came over tea, and was mischievously prompted by Russell? And that it was stated only when the paper was read—too late for the printed list, and unnoticed by the secretary?

In any case, Popper's notes for the meeting show how carefully he worked out his address to the club. What could be an initial outlining of his thoughts begins with the thesis: "We are students of problems using rational methods. These are real problems . . . not problems of language or linguistic puzzles."

The next stage appears to be a structure, headed "Methods in Philosophy:"

   I. Why I had to choose this subject
  II. Remarks on the History of Philosophical Method
 III. Appreciation and Criticism of the linguistic method in Philosophy
  IV. Some Theses on Philosophy and Method

After that comes a crowded page of text set out in columns and crawling round the edges. It contains the observation: "Philosophy got lost in preliminaries to preliminaries. Frankly, if this is philosophy, then I am not interested in it." By the next page the thinking process is obviously over and the speech itself is in view. We are in a position to hear Karl Popper's *ipsissima verba*.

"I was invited to open a discussion on some philosophical puzzle," the address would start, before going on to dissect "puzzle:" "The method of linguistic analysis of pseudo-problems. Problems disappear. Combined sometimes with a thesis about the nature of philosophy—an activity rather than a doctrine—the activity of cleaning up puzzles. Some kind of therapeutics, comparable to psycho-analysis."

At this point Popper attacks the invitation to address "some philosophical puzzle":

> All this is assumed in the invitation; and this is why I could not accept it. In other words, in your invitation is involved a fairly definite view of the nature of philosophy, and of philosophical method. Now, this is a view which I do not share. Thus the very fact that it was assumed more or less forced me to choose is [sic] as the subject of my talk.

It must have been here that Wittgenstein made his first intervention and battle was joined.

None of this points to any major involvement of Russell. All these notes are on LSE paper. It is unlikely that Popper wrote them between tea, dinner at King's with Braithwaite, and the meeting at 8:30. So the most plausible explanation is that Popper did discuss his paper over tea, Russell gave the arguments his backing, and Popper, in his anxiety to strengthen his relationship with his hero, perhaps wanting to flatter him, exaggerated the conversation's importance.

With its blend of detailed argument and eagerness to please, Popper's letter was doubtless aimed at building a continuing relationship with the man whom he had once said should be named in the same breath as Hume and Kant. Now and later, Popper was to be disappointed by Russell's failure to reciprocate, and by the uneven character of their association.

Hiram McLendon claims that he saw his tutor, Russell, on Saturday afternoon, and that Russell said he was so appalled by the "barbaric reception" Popper had received that he had already written to Popper to apologize. Popper, he told McLendon, was "a man of greater learning and erudition than all of those upstarts taken together." But the archives show that nearly a month passed before Russell wrote the letter already quoted. And then there was no reference to a joint victory, nor any taking up of the philosophical points Popper made.

THIS ROUTINE MEETING of the Moral Science Club—one of seven on the term card—gives rise to a third mystery. Along with "Did Popper lie?" and "Did Russell put him up to it?" comes, "Was Popper more familiar with Wittgenstein's later work than he let on?" In H3 he seemed peculiarly well briefed. Although language was a lifelong fascination of Wittgenstein's, the image of philosophy as "therapy," an activity comparable to Freudian psychoanalysis, belonged to the later Wittgenstein; likewise the use of "puzzles"—and all the metaphors such as our philosophical problems being like linguistic cramps. Yet Popper later insisted that he had been ignorant of Wittgenstein II and that his target had been Wittgenstein I. It seems a curious admission—that he had been attacking an out-of-date target—and more like another essay in false modesty. He had "wondered" about Wittgenstein modifying his doctrine. It seems unlikely that he did not enquire.

WHILE POPPER WAS WRITING to Russell and still mulling over the fight ("It was not the Wittgenstein I expected to meet"), the object of his contemplation had returned to his reflections on philosophy. On the Sunday, in his coded diary, there is the first nod toward the evening forty-eight hours earlier: "One can say about those who mock linguistic observations in philosophy that they do not see that they themselves are enmeshed in deep conceptual confusions."

What of Wittgenstein's attitude to Popper himself once he had been face-to-face with him? There exists one telling piece of evidence. Soon after the H3 meeting, Wittgenstein had scrawled a note to Rush Rhees, a former student and a close friend, who translated *Philosophical Investigations* after Wittgenstein's death. Barely legible, it talks of "a lousy meeting . . . at which an ass, Dr. Popper, from London, talked more mushy rubbish than I've heard for a long time. I talked a lot as usual. . . ." Michael Nedo, who has an archivist's encyclopedic knowledge of all things Wittgensteinian, glosses the word "ass." It describes, he says, someone who acts without thinking—a reference to a German proverb: "The ox and ass do, men can promise." Or perhaps "ass" meant "too Ringstrasse to merit attention."

Rubbish or not, Wittgenstein apparently felt the need to reply to Popper's arguments at a meeting of the MSC three weeks later. "Prof. Wittgenstein's main aim," say the minutes, "was to correct some misunderstandings about philosophy as practised by the Cambridge school (i.e., by Wittgenstein himself)." And the minutes also record Wittgenstein's assertion that "the general form of a philosophical question is, 'I am in a muddle; I don't know my way.' "

There remains one other curiosity about Popper's version of events. This relates to his journey back to London on the day after the poker incident. In *Unended Quest* he describes how, sitting in the train, he listened to two young people discussing a review of *The Open Society* "in a leftish magazine" and asking "who was this Dr. Popper?" But what magazine was it? The bulk of notices appeared in January 1946. There was no review in the *New Statesman* in October; *Tribune* had reviewed the book in January. Hugh Trevor-Roper had dealt with it in *Polemic* in May. Could this "memory" of Popper's also be false?

23

All Shall Have Prizes

What a statement seems to imply to me, it doesn't to you. If you should ever live amongst foreign people for any length of time & be dependent on them you will understand my difficulty.

—WITTGENSTEIN

Popper's own philosophy of science had this element of paranoia in it. Because what he used to teach us is that the nearest thing to a true theory is one that hasn't betrayed you yet. Any proposition is bound to let you down finally, but we cling on to the ones that haven't let us down yet.

—STEPHEN TOULMIN

WHEN VISITED more than fifty years later, H3 was still a home to scholarly brilliance, shared between the Astronomer Royal Sir

Martin Rees and the economic historian Emma Rothschild, married to the Nobel Prize-winning economist Amartya Sen. Books, journals, papers crowded the walls and occupied every flat surface. The sitting room felt decently comfortable; the armchairs had a well-sat-on look. There is a small sofa (not the original—this had been sold for five pounds to a fellow don; Braithwaite encouraged the purchase with the disclosure that it had borne silent witness to Wittgenstein's assault on Popper). Nevertheless, H3 seemed too small to have crammed in such extraordinary intellects as on that one night in 1946, and too conventionally scholarly to have witnessed such passions.

Outside, Russell, Wittgenstein, and Popper would recognize one of the most beautiful townscapes in the world as being untouched by the passage of decades—though they might find it an effort to push through the hordes of tourists on their way down King's Parade, notice that the college now has visiting hours, and pause to look up at the ancient stained-glass windows of King's College Chapel, which were still stowed in the cellars of the Gibbs Building in the immediate aftermath of the war.

But, while the room and its view might be virtually unchanged, it is hard to imagine a similar debate raging today—in Cambridge or anywhere else. The poker incident was unique in that it arose from the coming together of two visitors from a now vanished Central European culture. The meeting took place in the exhausted aftermath of a desperate struggle for European democracy and just as a new and equally dangerous threat to that democracy was taking shape. On the big issues it was not enough to be right—passion was vital. Now that sense of intellectual urgency has dissipated. Tolerance, relativism, the postmodern refusal to commit, the cultural triumph of uncertainty—all these rule out a repeat of the pyrotechnics in H3. Perhaps, too, there is currently so much specialization, and so many movements and fissures within higher education, that the important questions have been lost.

Who won on 25 October 1946?

In new democracies and closed societies, *The Open Society* retains its freshness and relevance. It has now been translated into over thirty languages, and further editions are constantly planned. But in Britain and America, Popper is slowly being dropped from university syllabuses; his name is fading, if not yet forgotten. This, admittedly, is a penalty of success rather than the price of failure. Many of the political ideas which in 1946 seemed so radical and were so important have become received wisdom. The attacks on authoritarianism, dogma, and historical inevitability, the stress on tolerance, transparency, and debate, the embracing of trial-and-error, the distrust of certainty and the espousal of humility—these today are beyond challenge and so beyond debate. If a resurgence of communism, fascism, aggressive nationalism, or religious fundamentalism once again threatened the international order based on the open society, then Popper's works would have to be re-opened and their arguments relearned. As he insisted, the future is not reached on steel tracks laid down in the past.

As for *The Logic of Scientific Discovery*, this can stake a claim to having been the most important work of the twentieth century in the philosophy of science, though even Popper's most loyal followers now concede the complexities involved in formulating a robust criterion of falsifiability. Nevertheless, two other figures in this area have become, if anything, more fashionable—Paul Feyerabend, whose interest in the language of the philosophy of science was rather Wittgensteinian in approach, and Thomas Kuhn, who first coined the phrase "paradigm shift" to describe what happens when one scientific framework for viewing the world is displaced by a radically new one. And it remains curious that the London School of Economics, which more than any other institution was Popper's academic base, has no substantial memorial to him. His office has been converted into a lavatory. (However, New Zealand is not allowing Popper to slip into oblivion, with plans afoot in Christchurch to mark his life by naming a building or street after him—nonsmoking zones presumably.)

Wittgenstein's reputation among twentieth-century thinkers is, by contrast, unsurpassed. His characterization as a genius is unchallenged; he has joined the philosophical canon. A poll of professional philosophers in 1998 put him fifth in a list of those who had made the most important contributions to the subject, after Aristotle, Plato, Kant, and Nietzsche and ahead of Hume and Descartes. The gleam in the eye that was evident in his friends and followers has been passed down to subsequent generations; they pore over his texts like Talmudic scholars divining wisdom from the Torah.

Oddly, however, his intellectual legacy is as ambiguous as so much of his writing; its substance is as elusive as the meaning of his philosophical pronouncements. His harshest critics say that his impact has been like his analysis of philosophy itself: it has left everything as it was. He blew through the world of philosophy like a hurricane, but in his wake there has been a settling back down. He was an inspiration for the Vienna Circle and for logical positivism, but logical positivism has been discredited (with Popper's help). He was an important influence on the Oxford language philosophers, but their approach has gone out of fashion. A line can be traced between Wittgenstein and the postmodernists—but he would be appalled to be held responsible for them.

Some Wittgensteinian ideas have become givens. Truth tables have become an indispensable tool of formal logic. The tautological nature of logic and mathematics is broadly acknowledged. "Meaning is use" has proved an enduring slogan: words have the meaning we assign to them. Language—like all rule-governed activity—is grounded in our practices, our habits, our way of life. But the majority of philosophers remain unconvinced that, in releasing us from the delusion that language mirrors the world, Wittgenstein has extricated us from all our problems. His emancipatory project has freed us from certain language-based confusions. Nevertheless, it is unclear that all our philosophical problems arise solely from our use of language. Whether we have good reason to believe that the sun will rise tomorrow seems to be a problem beyond language itself. So professional philosophers continue to grapple with such issues as the mysteries of consciousness and the relationship between the mind and the body—they do not believe that these can be solved by linguistic analysis. If Wittgenstein demonstrated that there were puzzles, most philosophers do not believe that he showed that there were *only* puzzles. Popper, fighting the problem corner, might see that as a partial victory—though of course he would settle for nothing less than unconditional surrender.

Of the great figures in twentieth-century philosophy, only a very few have given their names to those who follow in their path. Popper and Wittgenstein are two. In the philosophical lexicon there is no room for Russellians or Mooreians, Braithwaitians or Broadians, Schlickians or Carnapians. That one can be identified in academia as a Popperian or a Wittgensteinian is a testament to the originality of these philosophers' ideas and the power of their personalities. Those extraordinary qualities were on display in H3. The thrust of the poker becomes a symbol of the two men's unremitting zeal in their search for the right answers to the big questions.

And what of the sine qua non of this story? The happenings in H3 might be clearer, but the fate of the poker remains a total mystery. Many have searched for it in vain. According to one report, Richard Braithwaite disposed of it—to put an end to the prying of academics and journalists.

Popper in the 1960s. "He felt he had won." But not everybody agreed.

Chronology

26 APRIL 1889: Ludwig Josef Johann Wittgenstein born, eighth and last child of Karl Wittgenstein, millionaire industrialist and steel magnate, and Leopoldine, née Kalmus.

28 JULY 1902: Karl Raimund Popper born, third and last child of Dr. Simon Popper, a well-to-do lawyer, and Jenny, née Schiff.

1903–6: Unable to go to the Gymnasium because of lack of Greek after being educated privately, Wittgenstein attends the Realschule at Linz, shares school corridors with Adolf Hitler (1904–5), reads Weininger's *Sex and Character*, Hertz's *Principles of Mechanics* and Boltzmann's *Populäre Schriften (Popular Writings)*.

OCTOBER 1906–MAY 1908: At the Technische Hochschule, Berlin, Wittgenstein studies mechanical engineering and begins his philosophical notebooks.

1908: Wittgenstein goes to Manchester to study aeronautics.

1908: Popper learns the three Rs with his first teacher, Emma Gold-berger.

1908–11: Wittgenstein, a research student at Manchester University, reads Russell's *Principia Mathematica* and Frege's *Grundgesetz (Basic Law)*.

1911: Wittgenstein draws up plans for a book on philosophy, goes to see Frege in Jena, goes to Cambridge (unannounced) to meet Russell.

1912: Wittgenstein writes his first manuscript, is admitted to Trinity College, Cambridge, attends G. E. Moore's lectures, reads William James's *Varieties of Religious Experience*, becomes an influential member of the Moral Science Club, is elected to the Apostles, holidays with David Pinsent in Iceland, and visits Frege in Jena.

1912: Popper (aged ten) goes for walks, arranged by the Monist Society, with the antinationalist and socialist Arthur Arndt. They discuss Marx and Darwin.

1913: Wittgenstein holidays in Norway with Pinsent. Gives "Notes on Logic" to Russell.

OCTOBER 1913: Death of his father, Karl, makes Wittgenstein personally rich. He moves to Norway with the intention to build a house, to study and write.

APRIL 1914: Wittgenstein begins building a house at Skjolden. Dictates notes on logic to Moore.

28 JUNE 1914: Returning from a walk with Arndt, Popper hears of the assassination of Archduke Franz Ferdinand in Sarajevo.

JUNE 1914: Wittgenstein returns to his family country house in Austria, the Hochreith.

JULY 1914: Wittgenstein gives 100,000 crowns to artists in need, including Rilke and Kokoschka.

7 AUGUST 1914: Wittgenstein joins the Austro-Hungarian army as a volunteer following the declaration of war against Russia. Later he continues philosophical reading with works by Tolstoy, Emerson, and Nietzsche; develops his picture theory of language; begins work on the *Tractatus*. His brother Paul loses his right arm fighting on the Russian front.

1915: Popper presents written arguments against war to his father.

1915: Wittgenstein is wounded in an explosion in the artillery workshop in Cracow where he is serving.

1916: At his own request Wittgenstein is posted to an artillery regiment in the front line in Galicia. He is decorated several times. He continues writing the *Tractatus*.

1916: Death of Emperor Franz Josef.

1917: In a "key year," Popper is kept away from school by glandular fever.

1917–18: Wittgenstein serves on the front line against Russia in Bukovina and then against Italy near Asiago. He is awarded the Distinguished Military Service Medal with Swords.

1918: Popper leaves school without taking the *Matura* final exams and so cannot continue to the University of Vienna proper but enrolls as a nonmatriculated student.

JULY 1918: Wittgenstein finishes the *Tractatus*, under its original German title of *Logisch-philosophische Abhandlung*.

NOVEMBER 1918: Wittgenstein is captured by the Italians. A prisoner of war in Italy until August 1919, he sends the text of the *Logisch-philosophische Abhandlung* (the *Tractatus*) to Russell and Frege. In December 1919 he meets Russell in The Hague to explain the text, which has proved impossible to publish.

NOVEMBER 1918: Following crippling military defeat, the Austrian Republic is created. Popper witnesses disaffected soldiers shooting at members of the provisional government on the declaration of the Republic. The "hunger years" begin for Austria.

SEPTEMBER 1919: Wittgenstein hands over his personal wealth to his brother and sisters. Believing he has exhausted his philosophical possibilities, he enrolls in teacher-training college.

1919–20: In a turbulent period, Popper leaves home to spare his father expense and lives in a student barracks. He flirts with communism, but rejects it on seeing "in one of the most important incidents of my life" some young socialist demonstrators shot by police. He works for Adler, hears Einstein lecture in Vienna, and tries to earn his living as a road-mender but has not the physical strength to continue.

1920: Russell writes an introduction to the *Tractatus* to help its publication, but it is rejected by Wittgenstein. Wittgenstein takes up his first post as a schoolteacher, in Trattenbach in lower Austria. Posts in Hassbach, Puchberg, and Otterthal follow in subsequent years, and in 1925 he composes the *Wörterbuch für Volksschulen*, a 5,700-word dictionary for schoolchildren.

1921: *Logisch-philosophische Abhandlung* is published in the last number of Wilhelm Ostwald's *Annalen der Naturphilosophie*, with Russell's introduction.

1922: An English publisher, Kegan Paul, agrees to publish the *Tractatus* in a dual-language edition. Moore has suggested the title *Tractatus Logico-Philosophicus* for the English translation. In November, Wittgenstein receives his first author's copy.

1922–24: Popper works as a cabinetmaker's apprentice, passes the *Matura* for university entrance as external candidate, studies at the Vienna Conservatoire, and becomes a member of the Society for Private Music Performances presided over by Arnold Schoenberg.

1923: Wittgenstein inquires about the possibility of completing his Cambridge B.A. degree and is advised to work for a Ph.D.

1924: The leader of the Vienna Circle, Moritz Schlick, makes his first contact with Wittgenstein, whose ideas and book have become the subject of intense interest in Vienna.

1924: Popper finishes his apprenticeship as a cabinetmaker, obtains a primary-school teacher's diploma, and works with disadvantaged children.

1925: Popper begins studying at the newly founded Pedagogic Institute in Vienna; meets his wife-to-be, Josefine Anna Henninger ("Hennie"). In court over allegations of negligence in connection with a youth's injury, he is acquitted.

26 APRIL 1926: Wittgenstein is in court over his treatment of a pupil — the Haidbauer case. He quits teaching and begins work as gardener. With Paul Engelmann, he designs the Kundmanngasse house for his sister Margarete Stonborough.

NOVEMBER 1926: The Linz conference of the Social Democratic Party introduces the idea of armed conflict into its politics.

FEBRUARY 1927: Wittgenstein starts thinking about philosophy again. He meets Moritz Schlick. In the summer of this year he meets Carnap, Feigl, and Waismann — members of the Vienna Circle — for discussions on Monday evenings.

15 JULY 1927: Eighty-five demonstrators are killed by police at an attempted burning of the law courts in Vienna over the acquittal of three right-wing Heimwehr Frontkämpfers who had shot a cripple and a child in a street battle with the left-wing Schutzbund. Popper and his future wife witness the shooting.

1928: Popper successfully submits his doctoral thesis in the history of music, philosophy, and psychology. His is awarded the highest grade.

1928: Wittgenstein hears L. E. J. Brouwer lecture on the foundations of mathematics in Vienna — this is a major stimulus to new work.

JANUARY 1929: Wittgenstein returns to Cambridge. He begins work on a series of volumes later published as *Philosophical Remarks*.

18 JUNE 1929: Awarded a Cambridge Ph.D. on the strength of the *Tractatus*, Wittgenstein receives a study grant from Trinity College.

JULY 1929: The text of Wittgenstein's paper "Some Remarks on Logical Form" is published. He returns to Vienna for the summer, and later for the Christmas holiday, as he would do whenever possible.

1929: Popper qualifies to teach mathematics and physics in lower secondary schools. He makes contact with members of the Vienna Circle, including Viktor Kraft and Herbert Feigl.

1929: Wittgenstein begins an abortive collaboration with Waismann. He gives his first lectures on problems of language, logic, and mathematics, marking the nascent expression of his new approach to philosophy, and is awarded a one-off grant to enable him to continue his researches.

DECEMBER 1929: Constitutional changes strengthen the Austrian presidency at the expense of other democratic and constitutional organs.

1930: Popper begins teaching in secondary school and marries Hennie. He is encouraged by Feigl to begin writing the book that will become *Logik der Forschung (The Logic of Scientific Discovery)*.

1930: Wittgenstein begins teaching at Cambridge and attending the Moral Science Club. He applies for and gains a five-year research fellowship at Trinity, and moves back into his prewar rooms in Whewell's Court.

1931: Wittgenstein returns to Norway, and works on what will be published as *Philosophical Grammar*. Shares his first "confession" with friends.

1932: Popper completes *Die beiden Grundprobleme der Erkenntnistheorie (The Two Fundamental Problems of the Theory of Knowledge)*. In summer he stays in the Tyrol with Carnap and Feigl.

1932: Wittgenstein begins the "Big Typescript," the beginnings of a book that he then dictated to a secretary during a summer holiday at the Hochreith in 1933. It was then subject to constant revision. Accused of monopolizing the Moral Science Club, he temporarily withdraws from active participation.

7 MARCH 1933: In Austria, Engelbert Dollfuss formally abolishes parliament, making Austria a clerical semifascist state.

1933–34: Wittgenstein's "Blue Book" and "Brown Book"—lectures and course notes dictated by Wittgenstein to his students. Copies were made, the first in 1933 in a blue wrapper and the second in 1934 in a brown, and circulated in samizdat style.

12 FEBRUARY 1934: In a "civil war," the Schutzbund begins a rising—crushed by the army. All Social Democrat institutions are abolished and many party members are arrested.

25 JULY 1934: Dollfuss is shot during a Nazi putsch. Kurt Schuschnigg takes over as leader of the Vaterländische Front (Patriotic Front).

SEPTEMBER 1934: Popper attends an international philosophy conference at Prague, effectively run by the Vienna Circle.

DECEMBER 1934: Following a contract dependent on shortening his manuscript, Popper publishes a truncated version of *Die beiden Grundprobleme der Erkenntnistheorie* as *Logik der Forschung*.

SEPTEMBER 1935: Wittgenstein visits the Soviet Union.

1935: Popper lectures to Karl Menger's mathematics colloquium in Vienna.

1935–36: Wittgenstein lectures on "Sense Data and Private Experience." With the end of his research fellowship, he returns to Norway and begins work on *Philosophical Investigations*.

1935–36: On unpaid leave from teaching, Popper makes two visits to England, where he lectures, reads papers, and attends meetings in

London, Cambridge, and Oxford. He meets Moore, Schrödinger, Ryle, Ayer, von Hayek, Gombrich, and Berlin, goes to a meeting of the Aristotelian Society to hear Russell, and presents a paper at the Cambridge Moral Science Club. He then goes on to Copenhagen and meets Bohr.

1936: Wittgenstein writes a second confession for distribution to family and friends.

1937: Wittgenstein returns to Norway; Part 1 of *Remarks on the Foundations of Mathematics* is written. He shares his second "confession" with friends in England.

1937: Popper declines temporary refugee status in Cambridge, offered by the Academic Assistance Council, and accepts a permanent post at Canterbury University College, Christchurch, New Zealand.

1938: In New Zealand, Popper begins work that will become *The Poverty of Historicism* and *The Open Society and Its Enemies*.

FEBRUARY 1938: Wittgenstein stays in Dublin, visiting Maurice O'Connor Drury.

12 MARCH 1938: Austria is united with Germany in the *Anschluss*. Austria becomes Ostmark, a province of Greater Germany. Wittgenstein is advised by Piero Sraffa not to go to Austria and decides to apply for British citizenship and for a Cambridge post. His family in Vienna recognize the consequences of the *Anschluss* for them as Jews.

APRIL 1938: Wittgenstein returns to Cambridge to teach. Notes from his lectures will be published posthumously as *Lectures and Conversations on Aesthetics, Psychology and Religious Belief.*

FEBRUARY 1939: Wittgenstein is elected Professor of Philosophy. His fellowship at Trinity is renewed, and he returns to his rooms in Whewell's Court. He resumes an active role in the Moral Science Club.

JUNE 1939: Wittgenstein receives a British passport and travels to New York, Vienna, and Berlin in an attempt to secure non-Jewish status for his family in Vienna.

AUGUST 1939: Berlin issues the first certificate on non-Jewish status for Wittgenstein's sisters, as *Mischlinge* of the first degree under the Nuremberg Laws.

SEPTEMBER 1939: Britain declares war on Germany.

FEBRUARY 1940: Berlin officials determine that the Nuremberg Laws do not apply to the descendants of Hermann Christian Wittgenstein.

1941–44: Wittgenstein works as a hospital orderly in London and as a laboratory assistant in Newcastle upon Tyne.

MARCH 1944: Wittgenstein returns to Cambridge to resume his chair and to write and lecture. He becomes chairman of the Moral Science Club in succession to Moore. He spends from spring to autumn in Swansea with Rush Rhees.

1943: Popper finishes *The Open Society and Its Enemies*.

1944: Popper publishes *The Poverty of Historicism* in Hayek's *Economica*.

1945: Popper publishes *The Open Society and Its Enemies* in Britain. He is offered the Readership in Logic and Scientific Method at the London School of Economics, London University.

1946: Popper arrives in England on 6 January. Later in the year he is granted British citizenship.

25 OCTOBER 1946: Confrontation in H3 between Wittgenstein and Popper.

1947: Wittgenstein resigns his Cambridge professorship.

1947–48: Wittgenstein moves to Ireland and works on *Remarks on the Philosophy of Psychology*.

1949: Wittgenstein returns to England briefly, works on *On Certainty*, and visits Vienna to see his cancer-stricken sister Hermine. He visits Norman Malcolm in the United States. On his return to England he is himself diagnosed as having cancer.

1949: Popper becomes Professor of Logic and Scientific Method at the LSE.

1950: Wittgenstein writes *Remarks on Colour*, visits Norway for the last time, and moves into Dr. Bevan's house in Cambridge.

1950: Popper visits the United States to deliver the William James lectures at Harvard, and also meets Einstein at Princeton. He is able to buy a house in Penn, Buckinghamshire.

1951–53: Popper begins work on *Postscript: Twenty Years After*.

1951: Wittgenstein does further work on *On Certainty* but on 29 April dies in Cambridge at the home of Dr. Bevan.

1959: Popper's first publication in English of *The Logic of Scientific Discovery*.

1961: Popper attends the conference of the German Sociology Association in Tübingen with Adorno and delivers the Herbert Spencer memorial lectures in Oxford.

1962–63: Popper publishes *Conjectures and Refutations* and, at Schilpp's request, begins his autobiography for *The Library of Living Philosophers*.

1965: Popper is knighted.

1969: Popper gives up full-time teaching at London University to concentrate on writing and lecturing.

1972: Popper publishes *Objective Knowledge*, in which he elaborates his theory of the objective mind through Worlds One, Two, and Three.

1974: Two volumes of *The Library of Living Philosophers* are devoted to Popper and his thought—volume one contains his autobiography, later published separately as *Unended Quest.*

1983: A Popper symposium is held in Vienna.

1985: On the death of his wife, Popper moves from Penn to Kenley in south London.

1989: Popper delivers a public lecture at the LSE: "Towards an Evolutionary Theory of Knowledge."

1990: Popper publishes *A World of Propensities.*

1994: Popper dies on 17 September.

Appendix

Times Literary Supplement Letters

Between February and March 1998, the London *Times Literary Supplement* correspondence pages carried a series of seven letters reigniting the debate over what happened in H3 on 25 October 1946 when Karl Popper confronted Ludwig Wittgenstein. The series was sparked off by a letter from a Wittgenstein loyalist, Professor Peter Geach, fuming over a recent memoir of Karl Popper that rehashed Popper's version of the affray. Four of the letters are between Geach and the author of the memoir, Professor John Watkins. The remaining three correspondents weighed in with their testimony. The letters demonstrated both the passions the affair still roused and the conflicting memories of those present at the historic meeting.

1. Wittgenstein and Karl Popper

Sir,—In the Proceedings of the British Academy (1996), there has recently been repeated an old story of Karl Popper's about how

Wittgenstein threatened Popper with a poker, which he flung down when confronted by Popper, and stormed out of the room. Wittgenstein's surviving friends must often endure nasty old stories told about him, and protests have little effect; but in this case some response seems called for, lest the story should now seem to have scholarly accreditation.

At a meeting in London, Ontario, in 1976, this same story was rather irrelevantly put across by the speaker. The late Casimir Lewy and I, who had both been at the Cambridge Moral Science Club meeting to which Popper's story relates, made an immediate protest: Lewy called the story a complete fabrication, and I said, "Popper is a liar." Shortly before his death, Karl Popper wrote to me asking for an apology; I did not and will not apologize.

Peter Geach, 3 Richmond Road, Cambridge.
February 13, 1998

2. *Wittgenstein, Popper, and the Poker*

Sir,—As the person who wrote the memoir of Karl Popper published in the latest Proceedings of the British Academy, I may say that, pace Peter Geach (Letters, February 13), I did not simply repeat "an old story of Karl Popper's." I did, of course, make use of Popper's account of that meeting in *Unended Quest*, but I also looked into a good deal of other evidence. My account indicated various identifiable mistakes in his as well as other people's accounts and included some fresh material. Here are its main points, numbered in the hope that Professor Geach will be good enough to indicate precisely what he considers false.

1. The invitation from the secretary of the Moral Sciences Club contained the words "a few opening remarks stating some philosophical puzzle."

2. Popper's talk took off from the secretary's invitation.

3. He offered several examples of what he claimed were not mere philosophical puzzles but genuine philosophical problems.

4. Wittgenstein, who chaired the meeting, frequently interrupted. They were sitting on either side of the hearth, and at some point Wittgenstein took up the poker and began gesticulating with it rather freely.

5. Wittgenstein told Popper that he was confusing the issues.

6. Bertrand Russell called out to Wittgenstein that it was he who was confusing the issues.

7. One of Popper's examples of a genuine philosophical problem concerned the validity of moral principles. Asked for an example of a moral principle, Popper said something like "One ought not to threaten visiting lecturers with pokers," which caused laughter.

8. Wittgenstein left the meeting before it ended, slamming the door behind him.

The above points, except for (5) and (6), are also in Popper's account. Now to a disputed matter: Popper reported that it was to Wittgenstein that he said "One ought not to threaten visiting lecturers with pokers," and I went along with that. Geach recently wrote to me that Popper said it in the discussion after Wittgenstein had left. I withhold judgment on that. However, assume for argument's sake that Geach is right; would that justify talk of a "complete fabrication," or even talk of a small fabrication? The obvious explanation would be, not that Popper wilfully twisted this detail, but that his memory tricked him over it.

In conclusion, let me repay Geach's information with some information for him. At the meeting there was an American, Hiram McLendon, later a professor of philosophy at New York University, who was studying under Russell. The following afternoon he went to see Russell who (he said) spoke these words to him: "Popper is a man of greater learning and erudition than all of those upstarts taken together; and he is a person of great philosophical competence. The conduct of the vocal members of the Club toward him was shameful. In fact, so shameful was it that I have already this day written to Professor Popper a personal letter of apology for the barbaric reception given him here." Peter Geach, according to his letters to me, was among the vocal members. Whether or not his conduct then was shameful, his present conduct in putting "Popper is a liar" into print, without even bothering to vouchsafe any details of the alleged falsity, certainly seems shameful to me.

John Watkins, 11 Erskine Hill, London NW11.

February 20, 1998

3. *Popper and the Poker*

Sir, — In composing an account of Karl Popper's visit to the Cambridge Moral Science Club, John Watkins (Letters, February 20) did not consult me, though he knew I was there, and indeed mentions my presence in his account. For many years, Popper and his friends have been telling a dramatic story about the meeting: that Popper, confronting a furious poker-wielding Wittgenstein, offered as an example of a moral principle "thou shalt not threaten a visiting speaker with a poker" — whereupon Wittgenstein threw down the poker and left the room, slamming the door. This story is false from beginning to end. Watkins now writes to the *TLS* affecting to treat as a mere matter of detail, as to which Popper's memory might play tricks, the question whether Popper cited the "moral principle" before or after Wittgenstein left the room. If somebody falsely says "John and Mary had a baby and then got married," he would not be very well defended by a friend who said his memory might have slipped as to whether the birth or the marriage came first.

Peter Geach, c/o Mrs. Gormally, 44 Bromley Street, London E1.

June 3, 1998

4. Popper and the Poker

Sir,—The meeting of the Moral Science Club addressed by Popper and chaired by Wittgenstein seems to be in danger of becoming the stuff not of history but of mythology. As I was present at the meeting and have a tolerably clear recollection of it, it may perhaps be worth adding my testimony to the account given by John Watkins. My recollection accords with Dr. Watkins's, save that I do not recall any intervention by Russell or by Peter Geach or by any other member of the "vocal minority" of Wittgenstein's supporters while he was there. Wittgenstein sat forward on the edge of his armchair. In the course of a heated discussion, he picked up the poker lying in the hearth and used it as a baton to emphasize his comments. Shortly afterwards, Popper made the remark cited in Watkins's letter. Wittgenstein was clearly annoyed at what he saw as an unduly frivolous remark and he left the room.

I cannot say anything about what may have been said by Russell to Professor McLendon on the following day, but I do not find the remarks attributed to Russell surprising. At this time, Russell found himself out of sympathy with the dominant tone of philosophy at Cambridge. He very rarely attended meetings of the Moral Science Club. To assert that Wittgenstein threatened Popper with a poker, as I understand Popper did, is a gross exaggeration of an incident which did not carry, and was not seen as carrying, any hint of threat. Popper's observation was seen, by most of us at least, as a joke. Equally, there can be no justification for characterizing this exaggeration as a "complete fabrication" or as a "lie."

John Vinelott, 22 Portland Road, London W11.

June 3, 1998

5. *Popper and the poker*

Sir,—I hoped that my challenge would prod Peter Geach into providing details of the alleged falsity of Popper's account.

But no; he repeats (Letters, March 6) his story about what Popper said, adding "false from beginning to end" instead of "complete fabrication." But there is a significant shift. Hitherto, Professor Geach's story had Popper saying that Wittgenstein threatened him with a poker. If Popper had said that, it would indeed have been a gross exaggeration, as John Vinelott says (Letters, March 6); but Popper's remark about not threatening visiting lecturers with pokers was, of course, a quip (Popper said "joke"), and Vinelott confirms that it was seen as such at the meeting. The prime item in Geach's unscholarly story about Popper was Peter Geach's fabrication (his language is infectious) and is rightly, if silently, dropped from his latest letter.

John Watkins, 11 Erskine Hill, London NW11.
March 13, 1998

6. *Popper and the poker*

Sir,—What Peter Geach actually said at the Wittgenstein conference was not "Popper is a liar," but "And Popper will be damned in hell for a liar." But no one's memory is perfect. Nor did anyone think worse of Geach when he himself walked out of a Moral Science Club meeting after uttering what seemed to be an imprecation in Polish. If not wholly admirable, this style of criticism is as legitimate as that which prompted Roger Teichmann's legendary minute, "Professor Emmet woke up."

No, the real question about Karl Popper's encounter with Wittgenstein is "what sort of person walks out of a meeting which he is supposed to be chairing?" Surely the moral is that those who wish to dominate discussions ought not to chair them, and vice versa? And, as the Club's president, I am glad to say that this rule has been followed for the past twelve years.

Timothy Smiley, Faculty of Philosophy,
University of Cambridge, Sidgwick Avenue, Cambridge.
March 13, 1998

7. *Popper and the poker*

Sir, — Since Peter Geach (Letters, February 13) is worried that the story of how Wittgenstein threatened Popper with a red-hot poker is now being given "scholarly accreditation" (by John Watkins), I want to testify as an eyewitness of the incident that that accreditation is well deserved. Wittgenstein got very excited by Popper's insistence that there are serious philosophical problems, not just puzzles, as Wittgenstein would have it.

After a bit of discussion, Wittgenstein took the red-hot poker out of the fire and waved it in front of Popper's face. Bertrand Russell, who was facing the fire, sitting between the two men, took his pipe out of his mouth and said in his high-pitched, scratchy voice: "Wittgenstein, put down that poker at once!" Wittgenstein obeyed and, after a short time, got up and stormed out of the room. Geach saw and heard all this as clearly as I did, and it is incomprehensible that, out of mistaken loyalty to Wittgenstein, he should now deny that the incident happened. It is, however, unfortunate that some people have embroidered the story to the effect that Wittgenstein actually "threatened" Popper with the poker. Wittgenstein did not threaten Popper; he was merely overexcited and bad-tempered at being contradicted.

Peter Munz, Department of History,
Victoria University, PO Box 600, Wellington.
March 27, 1998

Acknowledgments

For a short book we have accrued a long list of debts.

First, to our eyewitnesses who kindly gave their time to search their memories, offer us their recollections, and furnish us with background information: Peter Geach, Peter Gray-Lucas, Wasfi Hijab, Georg Kreisel, Peter Munz, Stephen Plaister, Stephen Toulmin, Sir John Vinelott, and Michael Wolff.

Second, to those who knew one or both of our two heroes, or who were in Cambridge in 1946, and offered us similarly invaluable help: Joseph Agassi, the late Lord Annan, Peter Baelz, Joan Bevan, Lord Dahrendorf, the late Dorothy Emmet, Anthony Flew, Ivor Grattan-Guinness, John Grey, Christopher Hindley, Jancis Long, Hugh McCann (who worked for forty-seven years in the King's College kitchens), David Miller, Alan Musgrave, Arne Petersen, David Rowse, the late Dennis Sciama, Antonia Sewell, Jeremy Shearmur, Peg Smythies, George Soros, and Maurice Wiles.

Third, to the experts in various fields who generously shared their expertise, and sometimes work in progress: Peter Conradi, Sir Michael Dummett, David Earn, Elizabeth and Peter Fenwick, Sir Stuart Hampshire, Andrew Hodges, Allan Janik, Malachi Hacohen, Jutta Heibel, Liu Junning, Manfred Lube, Celia Male, Roland Pease, Charles Pigden, Friedrich Stadler, Barbara Suchy, Hans Veigl, and Sir Colin St. John Wilson.

A few must be identified for special recognition. John Watkins was the poker-research pioneer. Without his memoir of Popper, which reignited the controversy, we would not have written this book, and he provided us with many contacts and leads. Sadly, he died in 1999. John and Veronica Stonborough offered us generous hospitality and were equally openhanded with an array of useful facts and fascinating stories about the Wittgenstein family in general and Uncle Ludwig in particular. Joan Ripley put her reservations aside to talk to us about her father, Paul Wittgenstein. Stephen Toulmin responded unfailingly and always fully to our many requests for information and instruction about Cambridge and philosophy, Wittgenstein and Popper, Pyrrho and Sextus Empiricus. We met Melitta and Raymond Mew for a number of lengthy sessions that always began at the lunch table to ease the rigors of the journey to East Croydon. They read every sentence of two drafts, and in sharing their knowledge of and commitment to Karl Popper they saved us from many errors of fact and insight, though this does not signify their agreement with our portrayal. Michael Nedo, who could justifiably claim to be the most informed person in the world about Wittgenstein, gave us willingly of his time and erudition, and offered us the full support of the archive to which he has dedicated his life.

While we would not want to guarantee that all the statements that appear in this book are true, we have a number of people to thank for pointing out where our original versions were false. The following read the whole or part of the draft manuscripts, and made many useful suggestions: Roger Crisp, Hannah Edmonds, Sam Eidinow, David Franklin,

Anthony Grayling, Malachi Hacohen (who shared with us his own detailed account of Popper's early years), Peter Mangold, David Miller, Adrian Moore, Michael Nedo, Zina Rohan, Joan Ripley, Friedrich Stadler, Barbara Suchy, Stephen Toulmin, Lord Tugendhat, Maurice Walsh, and Jenny Willis.

We would also like to express our gratitude to Hannah Edmonds, Ron Gerver, and Lawrence Gretton for their assistance with translations, and Esther Eidinow for carrying out interviews in the United States. David would like to thank Liz Mardall of the BBC World Service for granting him leave of absence. John would like to thank Peter and Margaret Schlatter for providing perfect working conditions during his visits to Switzerland.

We were helped by several libraries, institutions, and archives: the Austrian State Archives; the Bergier Commission in Zurich; the Brenner Archive in Innsbruck; the British Library; Cambridge University Library; CARA (the Council for Assisting Academic Refugees); the City of London Barbican Library; King's College, Cambridge; the London Library; the London School of Economics (whose archive staff were unfailingly helpful); the National Archives in the United States; Bodleian Library, Oxford University; the Popper Library in Klagenfurt, Austria; the Popper Archive at the Hoover Institution; the Public Records Office in Kew; the Reichsbank Records in Frankfurt; the Russell Archive at McMaster University in Hamilton, Ontario; Trinity College, Cambridge; the Wittgenstein Archive in Bergen, Norway (especially Alois Pichler and Øystein E. Hide); the Trustees of the Wittgenstein Copyright; and the Wittgenstein Archive in Cambridge.

Extracts from unpublished papers of Sir Karl Popper are reprinted by kind permission of the estate of Sir Karl Popper.

Lines from the poem by Julian Bell quoted on pages 191–94 are reprinted by kind permission of Mrs. Quentin Bell. The verse of Rabindranath Tagore is reprinted with the kind permission of Visva-Bharati.

It would be wrong not to include here an acknowledgment to the *Times Literary Supplement* for publishing the exchange of letters that introduced us to the mysterious affair of the poker in H3.

And finally, and most importantly, we would like to thank Bernhard Suchy and Michael Neher at DVA and Julian Loose at Faber for being intrigued enough with our poker obsession to take it on, and our copy editor, Bob Davenport at Faber, and Julia Serebrinsky at Ecco/HarperCollins for their eagle-eyed attention to the text.

Photo acknowledgements: The pictures of Wittgenstein (ii and 23), King's College (xii), Braithwaite (63), Wittgenstein and his sisters (84), the hall of the Alleegasse (88), Moritz Schlick (145), the Kundmanngasse radiator (219), and Wittgenstein with Ben Richards (272) are reproduced with the kind permission of Michael Nedo and the Wittgenstein Archive. The pictures of Popper alone (ii, 28, and 294), Popper and his sisters (85), the Poppers' apartment (91), Popper with his class (109), and Karl with Hennie (273) are reproduced with the permission of Melitta Mew. Thanks are due to the Popper Archive for the MSC Term Card (29) and to the Russell Archive for the photograph of Bertrand Russell (50). The declaration of Hermine's assets (119) is from the Austrian State Archive. The photograph of Hitler (118) is from Ullstein Bilderdienst Berlin. The photograph of Peter Munz (13) was provided by Professor Munz himself. The photographs of Popper's cabinet (220) and of a poker in the hearth (277), are by John Eidinow.

Sources

The following books were particularly useful to us and are recommended to anyone interested in following up the topics covered in this story: Malachi Hacohen, *Popper: The Formative Years 1902–45* (Cambridge: Cambridge University Press, 2000); Ray Monk, *Wittgenstein: The Duty of Genius* (London: Cape, 1990); Michael Nedo and Michele Ranchetti, *Wittgenstein: Sein Leben in Bildern und Texten* (Frankfurt am Main: Suhrkamp Verlag, 1983); Friedrich Stadler, *The Vienna Circle: Studies in the Origins, Development, and Influence of Logical Empiricism* (Vienna and New York: Springer, 2000).

Our other sources and reference works were:

Agassi, Joseph. *A Philosopher's Apprentice* (Amsterdam: Rodopi, 1993).

Ayer, A. J. *Russell* (London: Fontana, 1972).

———. *Part of My Life* (London: Collins, 1977).

———. *More of My Life* (London: Collins, 1984).

————. *Wittgenstein* (London: Weidenfeld & Nicolson, 1985).

Bambrough, Renford, ed. *Plato, Popper and Politics* (Cambridge: Heffer, 1967).

Bartley, William Warren III. *Wittgenstein* (London: Cresset, 1988).

Beller, Steven. *Vienna and the Jews* (Cambridge: Cambridge University Press, 1989).

Bernhard, Thomas. *Wittgenstein's Nephew*, trans. David McLintock (London: Quartet, 1986); originally published as *Wittgensteins Neffe* (Frankfurt am Main: Suhrkamp Verlag, 1982).

Bett, Richard. *Sextus Empiricus against the Ethicists* (Oxford: Oxford University Press, 1997).

Bouwsma, O. K. *Wittgenstein Conversations*, ed. J. L. Craft and Ronald E. Hustwit (Indianapolis: Hackett, 1986).

Braithwaite, Richard. *Scientific Explanation* (Cambridge: Cambridge University Press, 1953).

Broad, C. *The Mind and Its Place in Nature* (London: Kegan Paul, 1925).

Carpenter, Humphrey. *The Envy of the World* (London: Weidenfeld & Nicolson, 1996).

Cartwright, N., et al. *Otto Neurath: Philosophy between Science and Politics* (Cambridge: Cambridge University Press, 1996).

Clark, Ronald. *The Life of Bertrand Russell* (London: Weidenfeld & Nicolson, 1975).

Cohen, R., Feyerabend, P., and Wartofsky, M., eds. *Essays in Memory of Imre Lakatos* (Dordrecht: Reidel, 1976).

D'Agostino, F., and Jarvie, I., eds. *Freedom and Rationality: Essays in Honour of John Watkins* (Dordrecht and London: Kluwer, 1989).

Dahrendorf, Ralf. *History of the LSE* (Oxford: Oxford University Press, 1995).

Davis, Norbert. *The Adventures of Max Latin* (New York: The Mysterious Press, 1988).

Drury, Maurice. *The Danger of Words* (London: Routledge, 1973).

Eagleton, Terry. *Saints and Scholars* (London: Verso, 1987).

———. *Wittgenstein: The Terry Eagleton Script, the Derek Jarman Film* (London: BFI, 1993).

Emmet, Dorothy. *Philosophers and Friends* (Basingstoke: Macmillan, 1996).

Engelmann, Paul. *Letters from Ludwig Wittgenstein* (Oxford: Basil Blackwell, 1967).

Essler, W., et al. *Epistemology, Methodology, and Philosophy of Sciences: Essays in Honor of Carl G. Hempel* (Dordrecht: Reidel, 1985).

Ewing, Alfred. *The Definition of Good* (New York: Macmillan, 1947).

Feigl, Herbert. *Inquiries and Provocations*, ed. R. Cohen (Dordrecht and London: Reidel, 1981).

Feinberg, B., and Kasrils, R., eds. *Dear Bertrand Russell* (London: Allen & Unwin, 1969).

Fenwick, Peter and Elizabeth. *The Hidden Door* (London: Headline, 1997).

Fogelin, Robert. *Wittgenstein* (Boston and London: Routledge & Kegan Paul, 1980).

Fraenkel, Josef, ed. *The Jews of Austria* (London: Vallentine Mitchell, 1967).

Friedländer, Saul. *Nazi Germany and the Jews Vol. I* (London: Weidenfeld & Nicolson, 1997).

Gadol, Eugene, ed. *Rationality and Science: Memorial Volume for Moritz Schlick* (Vienna: Springer, 1982).

Gates, Barbara. *Victorian Suicide* (Princeton, NJ, and Guildford: Princeton University Press, 1988).

Geach, Peter, ed. *Wittgenstein's Lectures on Philosophical Psychology 1946–7* (Brighton: Harvester, 1988).

———. *Peter Geach, Philosophical Encounters*, ed. H. Lewis (Dordrecht: Kluwer, 1990).

Geier, Manfred. *Karl Popper* (Hamburg: Rowohlt, 1994).

Gellner, Ernest. *Language and Solitude* (Cambridge: Cambridge University Press, 1998).

Golland, Louise, McGuinness, Brian, and Sklar, Abe, eds. *Reminiscences of the Vienna Circle and the Mathematical Colloquium* (Dordrecht and London: Kluwer Academic, 1994).

Gombrich, Ernst. *The Visual Arts in Vienna circa 1900* (London: Austrian Cultural Institute, 1996).

Gormally, Luke, ed. *Moral Truth and Moral Tradition: Essays in Honour of Peter Geach and Elizabeth Anscombe* (Blackrock: Four Courts Press, 1994).

Grayling, Anthony. *Wittgenstein* (Oxford: Oxford University Press, 1988).

———. *Russell* (Oxford: Oxford University Press, 1996).

Gulick, Charles A. *Austria from Habsburg to Hitler* (Berkeley and Los Angeles: University of California Press, 1948).

Hahn, Hans. *Empiricism, Logic and Mathematics*, ed. B. McGuinness (Dordrecht and London: Reidel, 1980).

Hamann, Brigitte. *Hitler's Vienna* (New York and Oxford: Oxford University Press, 1999).

Hanfling, Oswald. *Wittgenstein's Later Philosophy* (Basingstoke: Macmillan, 1989).

Hayes, John. Introduction to *M. O'C. Drury: The Danger of Words and Writing* (Bristol: Thoemmes Press, 1996).

Hijab, Wasfi. Unpublished memoirs.

Hilberg, Raoul. *The Destruction of the European Jews* (New York and London: Holmes & Meier, 1985).

Hirschfeld, Leon. *Was nicht in Baedeker steht* (Munich: Piper, 1927).

Hodges, Andrew. *Alan Turing: The Enigma of Intelligence* (London: Unwin, 1985).

Ignatieff, Michael. *Isaiah Berlin* (London: Chattto & Windus, 1998).

Janik, Allan, and Toulmin, Stephen. *Wittgenstein's Vienna* (London: Weidenfeld & Nicolson, 1973).

Janik, Allan, and Veigl, Hans. *Wittgenstein in Vienna* (Vienna: Springer, 1998).

Jarvie, Ian, and Pralong, Sandra. *Popper's Open Society after Fifty Years* (London: Routledge, 1999).

Jones, Ernest. *The Life and Work of Sigmund Freud*, ed. Lionel Trilling and Steven Marcus (London: Hogarth, 1962).

Josipovici, Gabriel. *On Trust* (New Haven and London: Yale University Press, 1999).

Kenny, Anthony. *Wittgenstein* (Harmondsworth: Penguin, 1975).

Kiesewetter, Hubert. *Karl Popper: Leben und Werk*, (Eichstätt: Eigenverlag, 2001).

Kraft, Viktor. *The Vienna Circle: The Origin of Neo-Positivism* (New York: Philosophical Library, 1953).

Kripalani, Krishna. *Tagore: A Biography* (London: Oxford University Press, 1962).

Leavis, F. R. *The Critic as Anti-Philosopher: Essays and Papers*, ed. G. Singh (London: Chatto & Windus, 1982).

Levinson, Paul. *In Pursuit of Truth* (Brighton: Harvester, 1982).

Levy, Paul. *G. E. Moore and the Cambridge Apostles* (London: Weidenfeld and Nicolson, 1979).

McGuinness, Brian, ed. *Wittgenstein and His Times* (Oxford: Blackwell, 1982).

———. ed. *Moritz Schlick* (Dordrecht: Reidel, 1985).

———. *Wittgenstein: A Life* (London: Duckworth, 1988).

McLendon, Hiram. Unpublished memoir of Bertrand Russell.

Magee, Bryan. *Popper* (London: Fontana, 1973).

———. *Confessions of a Philosopher* (London: Phoenix, 1997).

Malcolm, Norman. *Ludwig Wittgenstein: A Memoir* (London: Oxford University Press, 1958).

——. *Wittgenstein: A Religious Point of View* (London: Routledge, 1993).

Menger, Karl. *Morality, Decision and Social Organization* (Dordrecht: Reidel, 1974),

Minois, Georges. *History of Suicide* (Baltimore and London: Johns Hopkins University Press, 1999).

Monk, Ray. *Bertrand Russell: The Spirit of Solitude* (London: Cape, 1996).

Moore, G. E. *Principia Ethica* (Cambridge: Cambridge University Press, 1903).

——. *Commonplace Book*, ed. C. Lewy (London: Allen & Unwin, 1962).

Munz, Peter. *Our Knowledge of the Growth of Knowledge* (London: Routledge & Kegan Paul, 1985),

Musil, Robert. *Young Törless* (London: Panther Books, 1971).

Nemeth, E., and Stadler, F. *Encyclopedia and Utopia: The Life and Work of Otto Neurath* (Dordrecht. London: Kluwer, 1996).

Neurath, M., and Cohen, R., eds. *Otto Neurath: Empiricism and Sociology* (Dordrecht: Reidel, 1973).

Nicolson, Nigel, ed. *The Letters of Virginia Woolf Vol. 6* (London: Hogarth, 1980).

Paperno, Irina. *Suicide as a Cultural Institution in Dostoyevsky's Russia* (Ithaca and London: Cornell University Press, 1997).

Perloff, Marjorie. *Wittgenstein's Ladder* (Chicago, University of Chicago Press, 1996).

Pevsner, Nikolaus. *Cambridgeshire* (Harmondsworth: Penguin, 1954).

Pinsent, David. *Portrait of Wittgenstein as a Young Man*, ed. G. H. von Wright (Oxford: Basil Blackwell, 1990).

Popper, Karl. *The Open Society and Its Enemies* (London: Routledge & Kegan Paul, 2 vols., 1945).

——. *The Poverty of Historicism* (London: Routledge, 1957).

——. *The Logic of Scientific Discovery* (London: Hutchinson, 1959).

————. *Conjectures and Refutations* (London: Routledge & Kegan Paul, 1963).

————. *A Pocket Popper*, ed. David Miller (London: Fontana, 1983).

————. *Unended Quest: An Intellectual Autobiography* (London: Flamingo, rev. edn, 1986).

————. *A World of Propensities* (Bristol: Thoemmes, 1990).

————. *In Search of a Better World* (London: Routledge, 1992).

————. Kyoto lecture transcript: "How I Became a Philosopher without Trying" (1992).

————. *Knowledge and the Body-Mind Problem*, ed. N. Notturno (London: Routledge, 1994).

————. *The Myth of the Framework*, ed. M. Notturno (London: Routledge, 1994).

————. *The Lesson of This Century* (London: Routledge, 1997).

Redpath, Theodore. *Ludwig Wittgenstein* (London: Duckworth, 1990).

Rhees, Rush. *Recollections of Wittgenstein* (Oxford: Oxford University Press, 1984).

Rogers, Ben. *A. J. Ayer: A Life* (London: Chatto & Windus, 1999).

Roth, Joseph. *The String of Pearls*, trans. Michael Hofmann (London: Granta, 1998).

Russell, Bertrand. *The Problems of Philosophy* (London: Oxford University Press, 1912).

————. *Introduction to Mathematical Philosophy* (London: Allen & Unwin, 1919).

————. *An Outline of Philosophy* (London: Allen & Unwin, 1927).

————. *A History of Western Philosophy* (New York: Simon & Schuster, 1945).

————. *Human Knowledge: Its Scope and Limits* (London: Allen & Unwin, 1948).

———. *Logic and Knowledge*, ed. R. C. March (London: Allen & Unwin, 1956).

———. *Portraits from Memory and Other Essays* (London: Allen & Unwin, 1956).

———. *My Philosophical Development* (London: Allen & Unwin, 1959).

———. *The Autobiography of Bertrand Russell* (London: Allen & Unwin, 1967–9).

———. *Russell on Ethics*, ed. Charles Pigden (London: Routledge, 1999).

Ryan, Alan. *Bertrand Russell: A Political Life* (London: Allen Lane, 1988).

Ryle, Gilbert. *Collected Papers* (Bristol: Thoemmes, 1990).

Sarkar, Sahotra. *The Legacy of the Vienna Circle* (New York and London: Garland, 1996).

Sheamur, J. *The Political Thought of Karl Popper* (London: Routledge, 1996).

Schilpp, P. A., ed. *The Library of Living Philosophers—Moore* (Chicago: Northwestern University Press, 1942).

———, ed. *The Library of Living Philosophers—Russell* (Chicago: Northwestern University Press, 1944).

———, ed. *The Library of Living Philosophers—Broad* (New York: Tudor, 1960).

———, ed. *The Library of Living Philosophers—Carnap* (London: Cambridge University Press, 1963).

———, ed. *The Library of Living Philosophers—Popper* (La Salle: Open Court, 1974).

———, ed. *The Library of Living Philosophers—Georg Henrik von Wright* (La Salle: Open Court, 1989).

Schlick, Moritz. *Philosophical Papers Vol. 2*, ed. Henk L. Mulder and Barbara F. B. van de Velde-Schlick (Dordrecht: Reidel, 1979).

Segar, Kenneth, and Warren, John, eds. *Austria in the Thirties* (Riverside, Cal.: Ariadne, 1991).

Skidelsky, Robert. *John Maynard Keynes: A Biography Vol. 2* (London: Macmillan, 1992).

Smith, Joan. *Schoenberg and His Circle* (New York and London: Schirmer, 1986).

Stadler, F., ed. *Scientific Philosophy* (Dordrecht and London: Kluwer, 1993).

Steed, Wickham. *The Habsburg Monarchy* (London: Constable, 1913).

Stern, Fritz. *Gold and Iron* (London: Allen & Unwin, 1977).

Toulmin, Stephen. *Cosmopolis* (Chicago: University of Chicago Press, 1990).

Tagore, Rabindranath. *Gitanjali* (London: Macmillan Ltd, 1913).

Uebel, Thomas, ed. *Rediscovering the Forgotten Vienna Circle* (Dordrecht and London: Kluwer, 1994).

Waismann, Friedrich. *Philosophical Papers*, ed. B. McGuinness (Dordrecht: Reidel, 1976).

———. *Wittgenstein and the Vienna Circle*, ed. B. McGuinness (Oxford: Blackwell, 1979).

Walter, Bruno. *Theme and Variations* (London: Hamish Hamilton, 1947).

Wilkinson, L. P. *A Century of King's* (Cambridge: King's College, 1980).

Wisdom, John. *Other Minds* (Oxford: Blackwell, 1952).

Wistrich, Robert. *The Jews of Vienna in the Age of Franz Josef* (Oxford: Oxford University Press, 1990).

Wittgenstein, Ludwig. *Tractatus Logico-Philosophicus*, trans. C. K. Ogden (Routledge & Kegan Paul, 1922); rev. trans. by D. Pears and B. McGuinness (London: Routledge, 1961).

———. *Wörterbuch für Volksschulen* (Vienna: Hölder-Pichler-Tempsky, 1926).

———. *Philosophical Investigations*, ed. E. Anscombe and G. H. von Wright (Oxford: Blackwell, 1953).

———. *Philosophical Remarks*, ed. Rush Rhees (Oxford: Blackwell, 1975).

———. *Culture and Value*, ed. G. H. von Wright (Oxford: Blackwell, 1980).

————. *Zettel*, ed. E. Anscombe and G. H. von Wright (Oxford: Blackwell, 1981).

————. *Philosophical Occasions 1912–1951*, ed. James Klagge and Alfred Nordmann (Cambridge: Hackett, 1993).

————. *Cambridge Letters*, ed. B. McGuinness and G. H. von Wright (Oxford: Blackwell, 1995).

Wood, Alan. *Bertrand Russell: The Passionate Sceptic* (London: Allen & Unwin, 1957).

Wood, O., and Pitcher, G. *Ryle* (London and Basingstoke: Macmillan, 1971).

Wuchterl, Kurt, and Hübner, Adolf. *Wittgenstein* (Hamburg: Rowohlt, 1979).

Periodicals/Articles

Biographical Memoirs of Fellows of the Royal Society, 43 (1997), David Miller, "Sir Karl Popper;" *Encounter*, January 1969; *The Guardian*, 19 September 1994; *Journal of Modern History*, 71 (March 1999), Malachi Hacohen, "Dilemmas of Cosmopolitanism: Karl Popper, Jewish Identity, and 'Central European Culture;' " *Journal for the Philosophy of Science*, 3 (1952); *Mind*, vol. 56, no. 222 (April 1947), and vol. 60, no. 239 (July 1951); *New Statesman*, 30 August 1999; *Polemic*, May 1946; *Proceedings of the Aristotelian Society*; supplementary vol. 20 (1946); *Proceedings of the British Academy*, 94 (1997); *Russell*, vol. 12, no. 1 (summer 1992); *Sunday Times*, 9 December 1945; *Tages-Anzeiger* (Switzerland), 1 April 2000; *The Times*, 25 and 26 October 1946, 18 November 1946, 2 May 1951; *Times Literary Supplement*, 24 August 1946, letters published in February and March 1998, and Steven Beller review of *Thinking with History* by Carl Schorske, 15 July 1999; *The Venture*, no. 5 (February 1930); *Vierteljahrshefte für Zeitgeschichte*, April 1998, John M. Steiner and Jobst Freiherr von Cornberg, "*Befreiungen von den antisemitischen Nürnberger Gesetzen.*"

Interviews/Correspondence

Joseph Agassi, Lord Annan, Joan Bevan, Peter Baelz, Peter Conradi, Lord Dahrendorf, Sir Michael Dummett, Dorothy Emmet, Peter and Elizabeth Fenwick, Anthony Flew, Peter Geach, Ivor Grattan-Guinness, Peter Gray-Lucas, John Grey, Malachi Hacohen, Sir Stuart Hampshire, Wasfi Hijab, Christopher Hindley, Andrew Hodges, Allan Janik, Georg Kreisel, Melitta and Raymond Mew, David Miller, Peter Munz, Alan Musgrave, Michael Nedo, Arne Petersen, Charles Pigden, Stephen Plaister, Joan Ripley, David Rowse, Dennis Sciama, Antonio Sewell, Jeremy Shearmur, Peg Smythies, Friedrich Stadler, John and Veronica Stonborough, Barbara Suchy, Stephen Toulmin, Hans Veigl, John Vinelott, John Watkins, Maurice Wiles, Sir Colin St. John Wilson, Michael Wolff

Archives

Austrian State Archive, Vienna; Bergier Commission, Zürich; Brenner Archive, Innsbruck; Cambridge University Library Archives; King's College Archives, Cambridge; National Archives (FBI and CIA), Maryland; Bodleian Library Oxford (Academic Assistance Council records); Popper Archive at the Hoover Institution; Popper Library, Klagenfurt; Public Record Office, London; Reichsbank Records, Frankfurt; Russell Archive, Hamilton, Ontario; Trinity College Archives, Cambridge; Wittgenstein Archive, Bergen; Wittgenstein Archive, Cambridge

Index

"LW" indicates Ludwig Wittgenstein, "KP" Karl Popper
and "MSC" Moral Science Club